THE AI REPUBLIC

The **AI**
REPUBLIC

BUILDING THE NEXUS BETWEEN
HUMANS AND INTELLIGENT AUTOMATION

TERENCE C.M. TSE Ph.D,
MARK ESPOSITO Ph.D,
and DANNY GOH

LIONCREST
PUBLISHING

THE AI REPUBLIC

Building the Nexus Between Humans and Intelligent Automation

ISBN   978-1-5445-0283-0 *Hardcover*

   978-1-5445-0282-3 *Paperback*

   978-1-5445-0284-7 *Ebook*

To my wife, Celine, and daughters, Clélia and Lucile. Thank you for putting up with me.

—TERENCE

To my father, Vincent, and to the loving memory of my mother, Carla. Thank you for everything you did for me.

—MARK

To my family and my son, Nice, who has been supporting my dreams. And especially my wife, Teruyo, who never questions but believes in me.

—DANNY

To all our friends and colleagues at Nexus FrontierTech, who have put in their utmost to make the company a success.

—ALL THREE OF US

CONTENTS

INTRODUCTION

In 2015, the *New York Times* published a quiz designed to illustrate the writing capabilities of artificial intelligence technology.[1] Can you tell whether a human or a computer algorithm wrote the following passages?

1. "A shallow magnitude 4.7 earthquake was reported Monday morning five miles from Westwood, California, according to the U.S. Geological Survey. The temblor occurred at 6:25 a.m. Pacific time at a depth of 5.0 miles."
2. "Apple's holiday earnings for 2014 were record shattering. The company earned an $18 billion profit on $74.6 billion in revenue. That profit was more than any company had ever earned in history."
3. "When I in dreams behold thy fairest shade
   Whose shade in dreams doth wake the sleeping morn
   The daytime shadow of my love betray'd
   Lends hideous night to dreaming's faded form."

4. "Benner had a good game at the plate for Hamilton A's-Forcini. Benner went 2-3, drove in one and scored one run. Benner singled in the third inning and doubled in the fifth inning."
5. "Kitty couldn't fall asleep for a long time. Her nerves were strained as two tight strings, and even a glass of hot wine, that Vronsky made her drink, did not help her. Lying in bed she kept going over and over that monstrous scene at the meadow."
6. "Tuesday was a great day for W. Roberts, as the junior pitcher threw a perfect game to carry Virginia to a 2-0 victory over George Washington at Davenport Field."
7. "I was laid out sideways on a soft American van seat, several young men still plying me with vodkas that I dutifully drank, because for a Russian it is impolite to refuse."
8. "In truth, I'd love to build some verse for you
   To churn such verse a billion times a day
   So type a new concept for me to chew
   I keep all waiting long, I hope you stay."

Here are the answers: 1. Computer, 2. Human, 3. Computer, 4. Computer, 5. Computer, 6. Computer, 7. Human, 8. Computer

THIS IS NOT BUSINESS AS USUAL

How did you do? Don't feel too badly if you failed miserably. Even the most experienced literary scholar would have trouble deciding which of the above examples was written by a flesh and blood human.

The quiz illustrates how much progress machines have made in doing the types of things humans can do. It's clear that machines can already compose text in a way that is almost indistinguishable from human writing. This simply isn't business as usual; there is something going on here that is more important than many technological breakthroughs of the past. We've never seen such ambiguity between humans and machines.

Or have a look at what modern computers can paint. ING and Microsoft, a global bank and a tech giant, respectively, have recently completed a project called *The Next Rembrandt*.[2] Drawing on the analysis of 346 of the artist's paintings, a team of engineers used facial recognition technologies to create an original painting of the same name based on the styles and geometric patterns in Rembrandt's portraits.[3]

While this is exciting, and maybe a bit alarming, it's important to remember that even if they can write decent poetry, machines cannot, and never will, completely replace humans. There are in fact many roles machines will never be able to fill. To understand why, we need a better understanding of how AI fits into our world, and how it's evolving.

JUST HOW INTELLIGENT IS ARTIFICIAL INTELLIGENCE?

People often want to know if computers are becoming more intelligent than humans. This is an understandable, but ultimately unartfully phrased question. The underlying fear behind this question is the idea that AI represents an extraordinary technological development that has the potential to be smarter than we are, to maybe even dominate us. This is how the media, futurists, and science fiction frequently love to portray advances in AI. We have been warned, for instance, that the only way for humans to survive is to merge with machines.[4]

Undoubtedly, we can no longer dismiss the need to work in close conjunction with machines. But how much of a "merger" is really required hinges on how you define intelligence. It is a basic fallacy to assume that human intelligence is something that AI can ever possibly hope to emulate. This isn't the goal or end game for AI technology, either. In truth, what machines do isn't anything close to intelligence: they merely excel in the computational space. In spite of terms such as "machine learning" and "neural network" (which will be discussed in greater detail later), they are only reproducing analytical functions of the human brain, but this is only one part of what our brains are capable of. While this logical and analytical space can be easily replicated, other parts of our brain cannot.

Computers are smart in the sense that they are the most efficient when it comes to complex calculations and raw data processing. Computers will always be better ("smarter") than we are at that. Even a basic handheld calculator can process numbers at a speed we could never hope to achieve. With enough time (and math ability), you could do the same task, but a calculator will complete it in a nanosecond. In that sense, even a very dumb calculator is much quicker and smarter than we are. But is that intelligence? No, that is speed and raw number-crunching.

Arguably, there are some very advanced computational functions that computers can do now that we would never be able to accomplish as humans. But that is still not intelligence. Machines are grossly incapable of doing these things on *their own*; setting up the computers to perform a specific task still requires a significant amount of human input. A computer can perform statistical analysis, like calculating standard deviations and p-values, with commendable expediency and accuracy, but it cannot understand the purpose of doing so. The analysis and the resulting output are only meaningful thanks to the humans designing the study, picking the methodology, deciding what data needs to be collected, and calibrating carefully the calculating process. The human is still in the driver's seat, making the big decisions that will guide them to the results.

It's also important to remember that there are many things that machines cannot do very well. If you want a computer to help you dress better, for example, you will be out of luck. Achieving a superior sense of style is a far more difficult, subjective task to complete. Your welfare will be much better served by asking your closest fashionable friend for help or by hiring a human personal shopper, if you can afford it.

To us, the term artificial intelligence is misleading at its core, because it assumes a level of intellect that isn't necessarily there. At this stage, AI has a long way to go before it becomes truly smarter than we are, and it may never get there. Yes, AI can do a lot of impressive things, but can it serve a bigger, broader purpose for our society?

THE CHALLENGES TO THE RISE OF AI

The rate of technological change is accelerating exponentially. How long did it take for the telephone to reach every house in America? It was invented in 1876, and it took more than fifty years for that technology to get into half of American households. It took radio thirty-eight years to reach 50 million listeners. How long did it take Facebook, which was invented in 2004, to reach 6 million users and then later 600 million users? About one and five years, respectively.[5] Launched on January 21, 2011, WeChat, a messaging, social media, and mobile payment app developed by China's tech giant Tencent, reached 50 million users within a year. The number of users exceeded 1 billion some thirty *quarters* later.[6] There is no question that technology is now penetrating the market at a breakneck pace.

That said, some technologies never fully take off or become globally distributed. In Latin America, people still line up outside a bank to perform functions that are completed electronically elsewhere in the world. There are large parts of the world where electronic banking technology may never penetrate due to factors of cost and convenience.

Just like e-banking technologies, the pervasiveness of AI is much lower than what we have been led to believe. Despite the fact that today's media is filled with news of AI and portrays its widespread adoption in societies and businesses, the reality is very different. The rapid and large-scale deployment of AI is mostly confined to the largest tech corporates. Most entities in other sectors have yet to even consider integrating AI into their operations.

There are several factors that are currently restraining the rise of AI—the biggest one being cost. It's currently very expensive to design and employ AI systems. Most countries and companies around the world can't afford this kind of tech and won't be able to for many years. Right now, the largest players in AI are the United States, Japan, and China. Europe, which by no means is technological backwater, is lagging further and further behind and still has a great deal to catch up in this space.[7]

From the perspective of companies, AI is not a shelf-ready product; you can't simply buy a system and plug-and-play it in your business. For AI to work effectively, it needs to be tailored to the individual needs of an organization. It's a substantive process of integration where software developers and architects meet managers to work together to address specific needs. And this is assuming that the business needs and project objectives have been properly identified. Very often, even that initial assessment or detection of needs demands a lot of work. On many occasions, companies realize the solution to their issue isn't AI at all; what they need is to digitalize their processes and operations or to upgrade to a more modern infrastructure. Companies need to do plenty of homework before they can begin to dip their toes in the AI pool.

Another factor hindering the widespread use of AI is the hoarding of technology by major firms. Most AI tech that is developed by smaller firms is almost immediately acquired by larger companies like Amazon and Google. Often, leading corporates don't acquire that tech with the intent of commercializing it and making it widely available. Instead, they use it to improve their own businesses.

That technology is now off the market and unavailable for anyone else to use. They become a small, elite group of players in the AI world that have the power and war chests to heavily invest in the technology. They send scouts to smaller companies and quickly buy out useful technology, which then slows down the spread of technology and prohibits collaborative work to improve tech.

Way back in 2011, Marc Andreessen of Netscape fame wrote an article called "Why Software is Eating the World" explaining how prominent startups are "building real, high growth, high margin, highly defensible businesses."[8] In the seven years that followed, these startups, now turned tech giants, have not only created such businesses but also increased their market dominance, often to the level associated with monopolies.[9] The result: they are choking smaller rivals with their superior competitive edge.[10] These days, with the same tech giants holding a lead in AI technologies, they will most likely monopolize AI technologies and continue to put constraints on growth and developments on any upcoming challengers.

THE ENEMY WITHIN

The slow-going adoption of AI means it won't be taking over the world, at least not anytime soon. The media enjoys stoking worries about the "Rise of the Machines" because it gets it plenty of attention and page views. But you don't need to lose sleep over this. After all, the real threats to society as we know it have been around forever. A huge disruptive scenario on a global level is far more likely to come about due to climate change, war, trade war, or political disruption.

Humans are a far greater threat to humanity than AI could ever hope to be. We should be far less worried about the threat of artificial intelligence and far more worried about the threat of human intelligence. In many ways, our fundamental human characteristics make us bad at making decisions and choosing a course of action. Our minds are tripped up by error and prejudice.[11] Adding advanced technology to a volatile social situation is just fuel to the fire. "It isn't artificial intelligence I'm worried about; it's human stupidity that concerns me," said Peter Diamandis, founder of the X Prize Foundation as well as co-founder of Singularity University.[12] Historian Yuval Harari, on the other hand, warned that "one thing that history teaches us is that we should never underestimate human stupidity."[13] In the end, it's not the technology itself that's going to make or break us. It's the way we humans implement and use it. The way we tackle ethical and moral implications and apply the tech matters far more than the technologies themselves.

AT HOME, IN THE REPUBLIC

AI's threat to humanity has been overplayed, but that doesn't mean it's not a major driving force in the technology and business world that is going to irrevocably change the way we live. Even though we may not know what our destination will end up being, we certainly know which directions we are going. And for sure, humans will be working hand-in-hand with machines in an ever-increasing number of activities. The "merger" of man and computers began long ago, the moment we accepted technologies into our lives.

We are using more and more digital devices in day-to-day activities. In fact, in many respects, people have not just welcomed the use of but *delegated responsibilities* to machines. Just ask yourself how many of us still memorize telephone numbers these days? How many people still resort to reading maps going from point A to point B? We would always prefer to be *guided* rather than having to figure out the ways ourselves. At the same time, we are becoming more inclined to hand over privacy in exchange for convenience, as we are too happy to take advantage of "free" apps and online services by paying for them in the form of data.

This trend will only go on with improving machine intelligence as it will be able to assume—and automate—more and more important roles that humans have played in the past (think driverless cars). An implication is that our next social and technological developments will be much less dictated by governments, with their power to control diminishing. Instead, our choices will be shaped and informed by the available technologies (created by companies, in particular the tech giants) and how the others use them.[14] The number of "likes" or "stars" in rating you get don't come from the state; they are given by other humans. Twitter is more powerful than tear gas.[15]

The fact that the choice of how to charge ahead into the future and even the fate of nations lies in how human beings interact with AI—and how human intelligence is best combined with artificial ones—leads us to think that AI is effectively creating a republic. The Oxford Dictionary defines "republic" as "a state in which supreme power is held by the people and their elected representatives, and which has an elected or nominated president rather than a monarch."[16] While we are less concerned with the second part of the definition, we strongly believe that with the rise of intelligent automation, the power of shaping the future is in the building of the nexus between humans and AI. Whether this will represent a force of good or bad really depends on how much we understand and how we use AI technologies. To us, now is the time to seek and deepen such understanding.

This book is written with this in mind. It tries to address

AI's huge transformative potential for businesses and society. It attempts to shed light on the reality of what AI technology can do and what it will be capable of in the future. It will offer explanations of the different terminology and jargon we use to talk about AI. We will also have a deeper discussion on the broader implications of AI's increasing presence in our lives.

There is a lot of conversation right now on how to reconcile the capabilities of people and machines. While this is an important conversation, it's not the purpose of this book. This book will skirt the moral and philosophical questions that surround AI and leave that for others to tackle. Instead, we would like to explain the context surrounding these conversations. We will discuss the historical legacy of AI technology, some of which has been around for many years. We will also discuss the triggers that cause this technology to accelerate in capability and use. We would like to help people recognize how and why they might employ AI in their own organization, or city, or country. Drawing on our experiences at Nexus FrontierTech with helping our clients to integrate AI into their business capabilities, we will suggest a few ideas as to how companies can start to put AI into their organizations.

AI is not the answer to all our problems. There are many situations where employing AI could be detrimental to a business. The goal is not to simply replace old for new but to examine how AI can help advance the trajectory of your products or services. It's a matter of figuring out the things that machines should do and could do better.

In the latter part of the book, we will look at what governments can and should do to take advantage of this technology, and the implications of doing so. Given that all three of us have children, we believe that it is a worthy exercise to examine how parents can help prepare the next generations to fit into the economy of the future.

AI is constantly evolving, and society is being forced to evolve along with it. The ultimate question to ask yourself as you read is this: what are the things you should be doing today to prepare for the world of tomorrow?

PART I

WHAT IS AI?

CHAPTER 1

ARTIFICIAL INTELLIGENCE: FRIEND OR FOE?

It's very important to demystify AI, particularly if we want to deepen our understanding of how the technology works and figure out what we can do with it. If you were going into an ongoing business partnership with someone, you would want to learn as much as possible about your potential long-term partner. The same can be said here: knowing what AI is and what it can actually do is the first step towards building a beneficial collaboration.

But so far, it appears that artificial intelligence has acquired a negative reputation, and many people are scared by the very idea of AI. Some see it as a looming science fiction villain, destined to destroy humanity. Others see it as a kind of economic bogeyman, poised to steal the jobs of hardworking, blue-collar everymen.

While these fears are rooted in reality, it is safe to say that AI is not an imminent threat to humanity. This is because there is effectively no real intelligence in an AI machine. They are simply smarter versions of dumb tools. They aren't capable of complex thought, much less the enslavement of humanity.

The fact that AI is just a new and clever tool leads us to prefer to view the technology as a potential driver for new economic opportunity. We see full-fledged and applied AI as a beacon for the creation of jobs that don't even exist yet. While we may lose some jobs in the process, they can be replaced with new and different positions that work in tandem with machines. Clearly in this case certain segments of the society will make enormous gains while others are inevitably going to lose out.

A simple parallel would be the evolution of the internet and its effect on the economy. Ask anyone on the street if the internet has created new work opportunities, and you will receive a resounding yes. Although it has altered the economic landscape in countless ways, including job losses (like NASA's "human computers" before the arrival of mainframes) and shrinkage of industries (such as physical travel agents), we have many clear examples of jobs that completely revolve around the internet. AI will likely follow a similar path as it becomes more embedded in our lives.

The conversation around AI's stealing jobs is usually charged with emotion and doesn't lead us anywhere productive. While the concern may be valid, the solution does not lie in limiting the development of AI. The real question we need to be asking is how do we make sure that the technology that is currently being developed can propel us towards new jobs or professional careers that don't yet exist?

AI-PHOBIA

On July 31, 2017, there was a stream of news reports that Facebook had decided to pull the plug on a set of AI-driven chatbots that had begun talking to each other in a strange language that only they understood. The two chatbots were instructed by Facebook to try and negotiate a trade with each other, but they quickly began to communicate in a language mostly incomprehensible to humans. The exchange between the two machines went like this:

Bob: i can i i everything else...

Alice: balls have zero to me to me to me to me to me to me to me to me to

Bob: you i everything else...

Alice: balls have a ball to me to me to me to me to me to me to me

Bob: i i can i i i everything else...

Alice: balls have a ball to me to me to me to me to me to me to me

Bob: i...

Alice: balls have zero to me to me to me to me to me to me to me to me to

Bob: you i i i i i everything else...

Alice: balls have o to me to me to me to me to me to me to me to me to

Bob: you i i i everything else...

Alice: balls have zero to me to me to me to me to me to me to me to me to[17]

The exchange was so unnerving that the company ended up shutting the experiment down.

Yet this is enough for media to produce lurid headlines describing how "shocked" and "surprised" Facebook AI researchers were forced to kill the two "creepy" bots, in an attempt to pursue clickbait and sensation.[18] In reality, even though the social network company did produce bots that chattered in garbled sentences, they actually weren't alarming or surprising. Nobody at the company panicked, and neither should we. In fact, other than the gibberish conversations, the experiment was considered to have produced useful results and went on as expected. There is nothing scary, shocking, or noteworthy even. It is just a regular scientific experiment conducted on a regular day. Shutting down a chatbot when it ceases to show a reasonable result is about as ominous as replacing a faulty light bulb.[19]

This is an example of how deeply ingrained the fear of autonomous technology is in humans. It's the narratives we hear and are exposed to daily that inadvertently build a lack of confidence around the true role of technology. As a direct result of this fear, movies and television series are littered with examples of technological dystopias, from *I, Robot* to *The Terminator*. This tendency is nested in a bit of cynicism about the future of humankind, supported by a powerful negative message around the inherent evil of robots. After all, scaremongering has always been commercially very viable in all businesses.

In addition to doomsday-sensationalizing articles posted by some (wannabe) journalists, as well as catchy book titles suggesting how to survive AI or that this technology represents humans' final invention, it is not hard to see why so many people are AI technophobic. Journalists and authors have also spawned the widespread idea that the invention of artificial superintelligence will trigger unfathomable changes to civilization as we know it.

AGI VERSUS ANI

If July 31, 2017, the day Facebook's experiment was shut down, was ultimately an unimportant date, the truly worrying one would be August 29, 1997. This is called Judgement Day in the science fiction series *The Terminator*. On this day, the artificial intelligence Skynet, activated on August 4 of the same year, became self-aware at 2:14 a.m. Eastern time and started a nuclear strike on the United States of America, Russia, and other places, killing 3 billion people. Human beings, in a panic, tried to disconnect it, but to no avail. The grand result was that humans lived in a machine dominated world—and many blockbuster film installments and a TV series. Culturally, the story has entered our psyche, somehow inciting our fear of out-of-control, super intelligent machines.

Over the past decades, with technologies developing

at breakneck speed, it has dawned on many of us that there will be a Judgement Day moment in which men and machines will converge. Popularized by the futurist Ray Kurzweil, "technological singularity" or simply "singularity" describes the idea that ordinary human thinking will someday be overtaken by artificially intelligent machines.[20] But to understand why Skynet is an unlikely scenario anytime in the near future, you need to understand a bit more about where AI technology is in the present.

There are two different types of artificial intelligence: artificial general intelligence (AGI) and artificial narrow intelligence (ANI). AGI is also sometimes described as "strong AI," whereas ANI is "weak AI."[21] Right now, all computer intelligent technologies are working solely with ANI, machines with narrow intelligence that can only serve a very well-defined purpose. These machines do the jobs that we designate them to do and nothing more.

AGI is the kind of AI that scares people. Like *Terminator*'s Skynet, it's self-aware, with free will and priorities not dictated by any human. Many people's minds automatically jump to AGI when they hear about AI, particularly when they read stories like the previous one about Facebook. That kind of AI is very, very far out of reach.

Some people would relax the assumptions by arguing

AGI is what happens when machines can determine their own goals or seek out objectives. However, just because a machine can achieve a goal, it doesn't mean it's intelligent in the AGI sense. A missile, for instance, can achieve its goal of hitting a target. Do we consider a missile intelligent? Not really. Most of the decision-making behind launching a missile is done by humans. They determine the target, decide to launch or not, and press the button. The missile's goal at that point is to hit the target. This is not AGI.

It is therefore strange when well-known figures make claims as if we are approaching real AGI. For instance, it was Elon Musk who said, "I think we should be very careful about artificial intelligence. If I had to guess at what our biggest existential threat is, it's probably that...with artificial intelligence we are summoning the demon. In all those stories where there's the guy with the pentagram and the holy water, it's like—yeah, he's sure he can control the demon. Doesn't work out."[22] The sentiment was echoed by Bill Gates when he said, "A few decades after that though the intelligence is strong enough to be a concern. I agree with Elon Musk and some others on this and don't understand why some people are not concerned."[23] Even the late Stephen Hawking once ominously said that AI could spell the end of the human race.[24] Musk subsequently went on to point out AI could be the "most likely" cause of a third world war[25] and robots would be able to do everything better than we do.[26]

While celebrities can get away with far-fetched statements like these, ordinary souls like us will have to live with the reality. People who work day in and day out in the AI business know that the reality of AI is far, far away from that point. Even Gates backtracked recently to say he would publicly disagree with Musk to say that there was no need to panic over threats of AI on human existence.[27]

A more likely risk is that humans mismanage this technology, not that the AI starts managing us. If, for example, a government is unable to recognize the threat of replacing labor with AI without then creating new jobs. Or, if they are unable to figure out a new tax system that taxes companies using machines instead of humans and, as a result, tax revenues go down. This could be detrimental to society, not because the machines are taking over, but because our governments are unprepared to deal with the ramifications of economic change. This is a mismanagement of one's own institution, policies, and emotions, not because of the technology itself.

AN ATMOSPHERE OF FEAR

In the United States after World War II, the predominant driving feeling among the population was joy or optimism. A baby boom helped rebuild a decimated society, because people finally believed the future would be okay. After the Vietnam War ended, a similar thing occurred with an atmosphere of recovery and a shared spirit of creating a new foundation. These predominant feelings orient public sentiments towards many different current events.

Unfortunately, the current pervasive ambiance is one of anger and fear. You can see this across the United States and Europe with the fractured Brexit vote, contentious elections, and the widespread movement towards populism and nationalism. Our global mood leads our society to dwell on the negative implications, no matter what the subject matter. It's no wonder that our underlying fears and feelings about technology are sublimated into television shows like *Black Mirror*. As mentioned above, news media's narratives around AI tend to opportunistically play on the general population's fears and emotions. They are manipulating our fear of the unknown to create what is now the predominant vision of AI in most people's eyes. Creating these dystopian views of technology might be profitable for some people, but they aren't necessarily detached from vested interest. Hence, we strongly believe that we should present a more uplifting—and transparent—view of AI technologies. We must cut through the hype.

THE CHINESE ROOM

When we have seminars, we like to tease people to see if they know the difference between the Internet of Things, Big Data, and AI. We feel there is a clear demarcation and we should spend time lecturing people on these differences. Understanding the technology is key to dispensing with false preconceptions about its capabilities or implications. In the case of AI, we will have to question how much intelligence there exists within. One of the best places to start is the so-called Chinese Room Experiment thought up by the philosopher John Searle.[28]

Picture this: you are alone in a room with just a Chinese to English dictionary to keep you company. We are outside the room, composing an essay in Chinese. We then slide the essay through a slit in the wall. We have no idea what is inside the room. You proceed to translate the essay from Chinese into English.

Assuming you manage to turn out a good translation and hand it back through the wall to us, should we consider the room to be intelligent? Can we say the room speaks Chinese? We might be tempted to. But when all the parts are laid bare, the answer is obviously no. It's the fact that you were in the room and could use the dictionary to translate that was useful. The room itself is not intelligent.

What this thought experiment shows is that if we can

create a process that convincingly imitates the presence of intelligence, people will believe it's there even if it's technically not. This is essentially what today's AI technologies do. It looks impressive, but AI doesn't understand the tasks it is completing. It wouldn't understand the content of the activities undertaken. It is, if anything, an extremely efficient machine with the ability to perform a single task to the best degree possible.

This is why the term artificial intelligence is misleading. Intelligence implies logic, understanding, and self-awareness, and we know machines possess none of those. If people can understand that AI at this juncture is merely an improvement in efficiency, then there will be less confusion and fear.

WHAT IS AI, REALLY?

In many respects, what we should be thinking about is "IA," not "AI." We should be treating AI as an intelligent assistant, intelligent algorithm, or intelligent asset, which, in turn, would give the notion a completely different feeling.[29] AI is merely here to provide extra assistance, extra computing power, and extra help in managing assets. It's just making a smarter, more efficient version of a dumb tool, not a smart tool that itself can solve the problems humans can't.

Machines cannot think, and they cannot solve problems

on their own. They can only answer specific questions. Ask a computer how it can help increase the revenue of your business, and you are unlikely to get a useful answer. The problem is too open-ended for a machine to generate a real solution.

The board game Go is supposedly the most strategic game humans have ever devised. When a computer, AlphaGo, beat the human world champion at Go in 2017, it wasn't because AlphaGo was a better player or able to hold a more strategic view. The computer won because it could play the same game again and again by itself. In fact, without having to rely on data from games humans had played before, it was able to remove the constraints of human knowledge. In other words, the machine won by being able to perform a huge amount of number-crunching. That's really it.

As humans, we have always been driven to improve our own sense of performance, and technology is the easiest way to do that. We often use technology to help us with predictions, things like weather and traffic forecasts, for example. In the past, these forecasts relied completely on labor-intensive calculations and estimates. Accuracy is often rather low, too. With improving computing power, availability of data, and AI to help us process more information, we've been able to improve accuracy to a level previously unattainable. AI is therefore a powerful technology that can process multiple variations and subsets of information in a way that humans cannot necessarily do, and it's able to "learn" from the information it's given. This type of learning is directed and designed to achieve higher accuracy depending on the machine's purpose.

This has opened up a lot of new opportunities. If you can forecast the weather more accurately, can you also forecast more abstract things, like fraudulent purchases on the internet? Sure, you can forecast fraud, if you can design an algorithm to ask the right questions. You start from a conceptual model, turn your question into computer programming, and then you execute it. Because the machine can learn from different iterations, the chances grow that the machine will autonomously find things out you may have not anticipated. Yet it is important to bear in mind that whatever it discovers will remain within the parameters of the direction you set the machine to go in. It's not going to suddenly quit searching for fraud and come up with the ability to diagnose cancer.

Following this logic, it is naïve to think that machines on their own will be able to fix large-scale social issues that plague our society. Instead, we need to envision how we can employ laser-focused AI in many different fields, like politics. It would be great to use AI to increase voter turnout, prevent voter fraud, and to more accurately predict outcomes in exit polls. At the same time, we should be aware that AI makes it even easier to create gerry-mandered voting districts—a powerful method of voter suppression.[30] AI's proper usage will all come down to how we decide to employ it to make society a better place.

WE SHOULDN'T FEAR PROGRESS

The telephone replaced the telegraph, the internet replaced the telephone, and technology and history continue to march forward hand in hand.

As humans, we fear what we don't understand. When we do this, we hamper our ability to take full advantage of new technologies. What we really need to deal with is the ignorance that feeds into this fear. We are still teaching children the same things we were decades ago. As a result, it's difficult to think outside traditional constraints, to reinvent things in ways that we have never done them before. Instead, we try to define the areas we will allow technology to inhabit.

Some of the challenges we face in multiple countries daily may find themselves elevated to a higher chance of solvability, should we manage to integrate human skills with laser-focused AI capabilities. Take for example the credit scoring system in the US. It calculates this score using a mathematical model where you analyze an individual's spending habits and current debt to attempt to calculate the viability that they will pay back a loan. Does this system work? Not really. It simplifies how we manage our money to an unrealistic degree. All it takes is going on a family vacation for three weeks and missing a payment to throw your entire credit score off.

If your credit score could consider more than the few dimensions currently used to calculate the score itself, supported by what technology can do better than humans, we would have an improved system. We eventually would be able to use the score in a better and more equitable manner. The same could be true if the right level of AI capability could be added to healthcare, medical prescriptions, or even detection of serious disease. Where humans find themselves measured against their own limitations, technology, if properly integrated, can lift those hurdles and propel us into more perfective solutions.

When it comes to reimagining the societal issues, AI can help. It can simplify or offer new perspectives we might fail to see otherwise.

It's a fact that banks tend to make more money when people spend money than when they save. As a result, banks often encourage their patrons to spend instead of saving, by providing lines of credit and loan incentives. AI can help people to manage their cash flow better. It can help consumers to foresee possible low funds and shortage of cash, achieve financial goals, and discipline them to know how much they're spending every day. Technology can help correct consumers' distorted mindsets and thus incentivize saving.

We can also use AI to improve our system of standardized testing. People assume that test performance is indicative of intelligence, but we know that there are many factors that influence testing results, beyond mastery of the material. Tests are biased towards certain populations, like kids from single-parent households, low-income neighborhoods, or kids with disabilities or attention disorders. These tests are not designed to compensate for those issues, and the number of "outliers" often exceeds what you would find in a normal distribution model.[31]

Although a test cannot factor in all these possibilities, AI may be able to help. It can illuminate blind spots and detect anomalies in test results. There's not a lot of difference between detecting fraud and detecting a kid with special needs. AI can help identify children that would benefit from alternative teaching styles. We can then match kids with their needs and skills to improve their educational journey rather than marginalize them and make them feel unfit.

The same could be held true for teachers. There are thousands of dedicated teachers who go the extra mile to help children with special needs or with disabilities. Many of those teachers are disempowered by the current standardized testing system. If we can support them with the right program/algorithm, we may be able to exponentiate their potential rather than limiting them today. Without this technological intervention, we will continue down the path of generating more inequality and denying opportunities to worthy children.

FACT AND FANTASY

Politicians love to advocate technology as an easy solution to many problems. It's important to be technologically literate enough to be able to tell the difference between facts and fantasy.

Consider the imminent disaster known as Brexit. Pro-Brexit politicians love to say that installing technology at the borders is going to solve all the potential custom issues the re-establishment of a hard border will raise. Truthfully, this is a fantasy. The politicians don't understand the technology and have no business making such an argument.

While AI can help to identify problems, it is no panacea. There are engineering problems, and then there are social problems. AI can help with the former but not the latter. Coming up with an engineering solution to a social problem is not the answer. Without solving the social issues first, the technology won't be able to do much.

Take the example of Philadelphia schools. Many students in this city were underperforming on standardized tests, which were based on material from the designated state textbooks. Children who had access to the textbooks did much better than those whose schools did not have enough textbooks for students. The suggestion was made then to build a digital system that would help redistribute the textbooks among the schools. If School A had too many history books and School B didn't have enough, they could just send some over. As it turned out, this wasn't practical either, because the data wasn't available. Nobody knew where all the textbooks were, and the database was rarely updated. Individual teachers would ultimately have to proactively track down copies of books. This relies on the goodwill of the teachers. And such goodwill would be stretched very thin if the teachers were forced to key the details of different books into the system.

The bottom line is that if you don't confront the social issues involved, no amount of technology is going to improve a situation like this. We can't solve social problems with engineering solutions.

ECONOMIC TRANSFORMATION, NOT ANNIHILATION

We need to reframe the AI debate to shift the conversation away from machines taking all our jobs and towards AI as a source for job creation.

Yes, at first glance, the predictions on job losses are staggering. A study suggests that 45 percent of the daily tasks currently done by humans could be automated if current trends continue.[32] In our previous book, we mentioned the case of a chief financial officer at an investment bank. He was given the task of reducing the size of his staff by 80 percent, as off-the-shelf digital technologies could be doing the jobs occupied by humans.[33] One of the most frequently cited reports from 2013 points out that 47 percent of jobs in the US will be lost to automation in the near future.[34] Yet this is an unlikely scenario—assuming the number of employees in 2017 in the US to be 125 million, that would mean 59 million people will become jobless.

Yet a recent study conducted by OECD suggests that these fears are somewhat overblown. It found that only 14 percent of jobs in OECD countries are "highly automatable"—that is, their probability of automation is 70 percent or higher.[35] In the past sixty years of automation, a majority of the 270 occupations examined in the US remain in existence, with only one being entirely eliminated: elevator operators.[36]

Automated away	1
Technologically obsolete	5
Demand fell	32
Still exist today	232

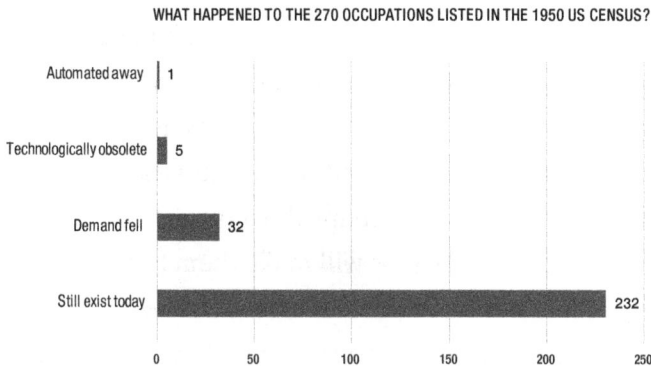

Figure 1. What happened to the 270 occupations listed in the 1950 US Census?

What's going on? Hollywood movies may have provided some hints. In the 1957 film *Desk Set*, the entire audience research department in a company is about to be replaced by a giant calculator. It is a relief to the staff, however, when they find out that the machine makes errors, and so they get to keep their jobs, learning to work alongside the calculator. Fast forward to the 2016 film *Hidden Figures*. The human "computers" at NASA are about to be replaced by the newly introduced IBM mainframe. The heroine, Dorothy Vaughan, decides to teach herself Fortran, a computer language, in order to stay on top of it. She ends up leading a team to ensure the technology performs according to plan.

These are not merely fantasies concocted by film studios (and in fact, *Hidden Figures* tells a true story). The fact is that while technology can do certain tasks better than humans, that doesn't mean it's going to eliminate humans completely. This is not like the combustion engine replacing horses. We are making room for technology to run alongside us. No doubt, certain jobs, especially those involving repetitive and routine actions, may succumb to automation for good, but the nonstandardized jobs are sticking around. Perhaps the automation of different aspects of these jobs will make them better. This is why estimates that 47 percent of jobs will be replaced by machines may be correct overall, but they don't actually mean that the 125 million jobs in the US will shrink to only 66 million. As academics, both Terence and Mark remember the arrival of Massive Open Online Courses or MOOCs. Until recently, it was widely believed that the rise of digital teaching tools would make us less relevant, or even superfluous, kick-starting the demise of our careers. But are we really going to see AI technology replace college instructors? No. Electronic courses are not taking away jobs. They have shown that human teachers can be made more effective with the use of digital tools. They are expanding the idea of what the instructor's job entails. They are complementary, not a substitute.

Ultimately, it is this combination of human interactions and computers that experts champion.[37] Social aspects matter in deployment of technologies. Additionally, certain jobs will always be done by humans. What may be surprising is that even work that requires no skills at all will not disappear any time soon. Housekeepers, for example, will in all likelihood never be replaced. Humans are just too good, efficient, and cheap at performing the variety of tasks necessary to deep clean hotel rooms than any one machine could ever be.[38]

O-RING THEORY

On January 28, 1986, the space shuttle *Challenger* exploded seventy-three seconds after its launch, killing all seven crew members onboard. A lot of analysis was done afterward to find out what went wrong. The fault came down to an inexpensive and seemingly inconsequential piece of rubber called an O-ring, a gasket that is meant to stop rocket fuel from leaking. As it happened, the night before the launch, Florida experienced unusually cold weather, which caused an O-ring to freeze and crack. The next day, the fuel leaked into other parts of the engine and caused the explosion.

This disaster led to what is now called the O-ring theory in economic development.[39] This theory suggests that a reliance on machines makes human skills *more* important, not less. Many of our current undertakings, from a space mission to producing goods and services, are composed of a series of interlocking steps, like the links in a chain. A variety of elements goes into these steps to generate the value of an activity including labor and capital, brain and physical power, exciting new ideas and boring repetition, technical mastery and intuitive judgment, perspiration and inspiration, adherence to rules, and the considered use of discretion, etc. Yet for the overall activity to work as expected, every one of the steps must be performed well. Therefore, if there is just one faulty piece in the chain, the whole process can break down. Each link must do its job for the chain to be complete, useful, and value-creating. In fact, if we were to make one step or link in a chain more robust and reliable, we are effectively improving the value of other links as the value of the entire chain goes up.[40]

Looking through this lens, automation does not necessarily make humans superfluous. Not in any fundamental way, at least. Instead, it increases the value of our skillsets. As AI and robots emerge, our expertise, problem-solving, judgment, and creativity are more important than ever. A recent study focusing on a Californian tech startup illustrates this point.[41] Despite the company's providing a technology-based service, it found itself to be growing so fast, with the computing systems getting larger and more complex, that it was constantly drafting in more humans to monitor, manage, and interpret the data. Here, the technologies are merely making the human skills more valuable than before. So for our purposes, for a chain to be robust, you need to make sure a human is doing the job of supervising the machines. Even if a chain were 100 percent machine driven and automated, it would still require humans to maintain it. In other words, full value needs to be placed on human intervention.

Let's not discount the importance of humans in anything. Any chain or process that creates value is going to require humans, and you cannot afford to undermine that. For machines to do the best job, you need humans to be doing the more important job of supervision.

COBOTICS AND COEXISTENCE

Do you remember how people once thought the ATM would lead to a drastic reduction of bank employees? There was a lot of fear around this notion that, ultimately, was unfounded. In the end, *more* jobs were created in the banking industry *after* the introduction of the ATM.[42] This fear of human obsolescence has always been present. For this reason, the conversation around AI needs to be geared towards the transformation of human jobs, not their disappearance.

What we are really losing is the tedium of a given job. At least half of the jobs that exist right now existed in a similar form one hundred years ago.[43] How much of your job is repetitive? If it's a lot, then it becomes a design issue. The problem is not the technology; it's that we are employing people in poorly designed jobs. The majority of jobs we are losing right now are manufacturing jobs to people overseas because of cheaper means of production. So as not to lose their jobs, the question should be how do we create new, better-designed jobs that coexist with technology?

That is where "cobotics" or simply "cobots" enter the picture. This term, which is gaining in popularity, describes humans working alongside robots to achieve common goals. Previously, humans were tasked with executing labor processes, which have now been taken over by machines. The person moves into either programming or supervising the process.

Think about manufacturing in the future. At some point, it will be entirely automated. In this case, those people that are displaced from the manufacturing floor will have new jobs either in management or performing maintenance on the machines.

Many people believe that self-driving trucks and cars will be the next big job displacement. The ramifications of such a shift are huge in terms of design. It will require rethinking major parts of our infrastructure. Yes, some of the 2 million long-haul truck drivers in the US may be losing the responsibility of driving actual trucks, but there are going to be many other jobs and opportunities that appear as we redesign a whole industry. For instance, shippers must find ways to protect unmanned cargo from theft, trucks will still need to refuel on trips across the country, and tires will have to be changed at times, something impossible for autonomous trucks to take on. Without the presence of drivers, support workers in the field play a vital role to ensure the vehicle remains in working order with maintenance checkups.[44] The problem, of course, is how to train up these drivers when most of them hold just a high school diploma and are the primary wage earners for their household.

We need to start thinking about what aspects of our jobs can be automated to make us better workers and free us up to doing other things. In the same way that advances like the internet and the ATM changed our world, AI will force us to clearly identify the valuable parts of our jobs as we offload the busy and unglamorous work.

CHAPTER 2

THE FOURTH INDUSTRIAL REVOLUTION

AI's rise to prominence didn't happen in a void. It is simply an overture in a much larger historical movement known as the Fourth Industrial Revolution, which is itself the result of hundreds of years of technological advancement.

The organizing premise of the various industrial revolutions is that technological evolutions happen incrementally. Throughout most of history, each time we have had a major technological breakthrough we also witnessed massive improvements in other related spheres: communication, transportation, or power. The First Industrial Revolution was mechanical: the invention of the steam engine, the creation of factories, and the harnessing of coal power. The second was scientific: the harnessing of electricity and mass production, which led to the rise of our consumerist society. The third revolution was all about computers, new ways of processing and sharing information, and the rise of the internet and globalization.

These previous revolutions were characterized by insularity. There was an initial technological leap and then a massive buy-in by the rest of society. Technological advancements eventually became general-purpose technology, the things we now take for granted, like electricity and the internet.

What is not to be taken for granted, however, are the many social benefits that arrived with each industrial revolution. When advancements in power, communication, transportation, and production were achieved, it led to more secure wages, the price of goods and services came down significantly, and the quality of life rose overall.[45]

There was great excitement when the World Economic Forum announced the arrival of the Fourth Industrial Revolution in 2016. Its essential statement was that what we are experiencing right now is fundamentally different from anything we've seen before.[46] The fourth movement is functioning slightly different than previous ones, in that it is currently happening concurrently with the Third

Industrial Revolution. We are still growing and improving computers and digital systems for sharing information, but at the same time, cyber-physical systems are exploding into being. This throws our chronology completely off, yet the potential for new and unforeseen social benefits—on top of increasing the social advancements we attained in previous industrial revolutions—will likely be greater than ever before.

BREAKING ALL THE RULES

In this new revolution, the technology being created not only disrupts our old ways of doing things, it creates entirely new ways for technology to exist in the world. Therefore, the Fourth Industrial Revolution is more of a foundational revolution, in that it is a revolution based on foundational technologies. Such technologies make it possible to create useful applications that enhance social progress and commercial pursuits. In that vein, a foundational revolution lays down the technological groundwork to facilitate the development of new inventions and discoveries for the same ends.

For example, the advent of APIs (Application Programming Interfaces) has made it possible for different internet and mobile software programs to communicate with each other. Without APIs, companies like Dropbox would never exist. We can also recognize APIs in action when we think about how Google Maps can be integrated into other websites and mobile apps as well as used on its own. It can be incorporated into other apps because of an API. The impact of a programmer's ability to use an API to integrate other products from other companies like Google into their own software cannot be overstated. The Google Maps API has helped thousands of companies to provide directions and locations without having to design, create, and implement their own mapping system, a time-consuming and complex task. This in turn frees up each company's bandwidth and resources to focus on developing new innovations and tools that no one has yet created. And just as important, without APIs, we would have a much slower and more fragmented experience on our internet browsers, desktop computers, mobile phones, and social networks. Productivity on a global level would never be anywhere as great as it is today. Technologies like APIs are what help define the Fourth Industrial Revolution as a foundational revolution.

Moore's Law was created by Gordon Moore, the co-founder of Intel. It states that technological performance doubles in capacity every two years. This has been true for several decades now, but modern technologies are beginning to increase exponentially faster than what Moore's Law dictated, and they defy commonly held beliefs about how technology advances.[47]

Instead of evolving linearly, these technologies are beginning to converge with one another. Entirely new fields and business models are being created, simply by combining elements of different technologies. The internet is a popular example of a convergent technology. While videos used to be watched on television, music played on the radio (or cassette player, or CD player, or even mp3 player), and poems were read in books, now these activities can be conducted in a single place—online. In the physical world, smartphones have been a convergent technology for more than a decade now. In one handheld device, you have the functionality of a desktop computer, a portable gaming system, a camera, a telephone, an mp3 player, and more.

Now convergent technology is also beginning to seep into the biological world, as new advancements literally alter our bodies. Think of personalized medicine and genome sequencing, for example, or gene editing. Diseases can now be treated based on genotypes, or even one day avoided altogether by altering your DNA (at the moment, gene editing has not been evaluated for its health risks and also presents an ethical dilemma). "Wearables" are another area of convergent technologies that are designed around the human body. Smart watches can track heart rate and other health indicators while also telling time. Other wearables, like virtual and augmented reality headsets (VR and AR), can help stroke victims and other patients retrain their brains during rehabilitation. In the workplace, VR headsets can also be used to design manufacturing processes, improve operator productivity, and reduce training time.[48]

As we can expect, some robotics technologies are now beginning to change the principles of our traditional labor market, and, by altering our historical assumptions of the purpose of work, we challenge the very meaning of human life. With many aspects of manual and intellectual labor being replaced, as discussed in the previous chapter, we as humans are being forced to redefine what it means to be engaged in the workforce and how work even fits into our overall life purpose.

AI, however, is the king of all convergent technologies. It incorporates so many of the various technologies that are currently at the forefront of the Fourth Industrial Revolution. It benefits from Big Data and helps automate decision making. It can also integrate into the Internet of Things, another foundational technology that functions as a platform for other technologies like near-field communications (NFC) to communicate with each other.[49] AI's relevance in IoT is its ability to take data as it is collected, analyze it, and decide if an action is needed (for instance, if an ambulance should be requested or if an alert message to a user needs to be sent out). In manufacturing, AI can also help with physical production in 3D printing with its ability to iterate and learn during cycles of production. All of these technologies bring together the tangible and intangible worlds in a unique way, with AI as a facilitator.

This is where the Fourth Industrial Revolution's power really lies, and if countries and companies adapt AI to tie together other technologies and understand the potential applications for social causes, they can begin to deploy technology in spaces like healthcare to save money and reach more people. The potential applications are endless.

Developments in the Fourth Industrial Revolution's foundation technology are not happening uniformly on a global basis. In some places, multiple industrial models are happening concurrently. For example, in India, manufacturing is a huge industry, but the country's production systems are nowhere near as advanced as in many parts of the world. Therefore, it is still representing many traits that we consider hallmarks of the Second Industrial Revolution. Yet in other spaces, like information technology (IT), digitalization has already entered the picture, and the landscape looks much more like what we consider the Third Industrial Revolution. At the same time, India also has its own space program, where the technology involved is so advanced that it's more characteristic of the Fourth Industrial Revolution. So the country is advancing in these different eras of technology concurrently.

In Rwanda, after the devastating civil war, the country managed to skip entire revolutions completely by fully digitalizing and jumping directly into the fourth. Similarly, former Soviet countries like Estonia, which didn't really exist until after the collapse of the USSR, have experienced very rapid technological penetration. In countries like this, you can see a digital technology like Skype completely overtake the entire telecom industry before it even had time to properly form. These societies are almost at an advantage because they don't have to reckon with replacing outdated foundational technologies.

In countries that have been able to skip directly into the Fourth Industrial Revolution, we see intriguing opportunities to build a foundation based on convergent technology. Countries in the African Union are an example of this potential. Because the continent did not participate in the earlier industrial revolutions, it is not beholden to older, limited infrastructure like telephone lines and coaxial cables. Though Africa is still affected by the impact of colonization and enslavement, countries can jump straight into wireless connectivity via satellites and mobile networks.

Africa will collectively have one of the world's largest workforces by 2030.[50] With an impressive workforce of this size, its countries can herald the foundational revolution: no ties to industrial artifacts, convergent technologies like AI and VR drive production and pro-ductivity, and everyone is digitally connected with access to data and information available at a very low cost. With foundational technology underpinning the economy like this, new inventions, innovation, and entrepreneurship will not be far behind.

SOCIALLY DRIVEN INNOVATION

As we mentioned in chapter 1, technology isn't a panacea to every problem. The simple fact is that we can do things faster now with help from technology. The way we choose to employ that technology, and the larger policies we create around it, is what will cause massive societal changes.

In addition, the benefits of making big shifts in produc-tion are self-evident. For example, replacing coal with gas was an easy choice because it was more efficient, it cost less, and it could be distributed more easily. The inter-net made communication much easier by shrinking the time it takes to receive a response. The tech in the Fourth Industrial Revolution doesn't just supplant something with more-powerful technology; it's also about the utility of the technology within social structures. For this reason, the Fourth Industrial Revolution is more socially oriented than the previous one.

The transportation app Uber is a useful example. It connects supply and demand between riders and drivers more efficiently than taxis while achieving several social benefits simultaneously: it keeps unnecessary cars off the road, makes city travel more affordable and accessible, and creates a source of income for those who in the third revolution never imagined they could ever use their own car to pay bills.[51]

One of the most exciting developments in the Fourth Industrial Revolution is the emergence of advanced bio-materials, called green chemistry. Green chemistry allows us to accomplish environmentally positive actions, like taking greenhouse gases like carbon dioxide and converting them from a waste product into a usable resource. For instance, chemists have found a way to use CO_2 as an industrial refrigerant. Known as transcritical CO_2, it can be used to reduce the climate impact of a building by 15 percent while also replacing chlorofluorocarbons and hydrofluorocarbons—environmentally damaging greenhouse gases that are currently used for refrigeration.

Along with green chemistry, we can use 3D printers to reproduce sections of coral reefs that have been destroyed by invasive human practices.[52] We can physically insert the byproduct of this technology into the ocean, and it will replace lost environments necessary for ocean biodiversity. We are using technology to answer an issue of social consciousness—the fact that we've lost 50 percent of the ocean's coral reefs. Humanity can use technology to fix some of the damage that previous technologies have caused to the environment.

Or consider plastics. There is a major conversation occurring right now that is about not just replacing one plastic with another but using less-harmful, green plastics. One idea is to use bacteria to break down plastics faster.[53] Other research involves using carbon dioxide and plant byproducts (like the leftover pulp from carrot juice) to make plastic.[54]

We have a social mandate that it is humanity's responsibility to improve our current waste disposal practices so that we can decrease pollution and ultimately improve our own health (because the fish we eat also consume the plastics we dispose of). New technology can help us achieve that goal and improve the planet in multiple ways.

The power of the Fourth Industrial Revolution is that it emboldens the social mandate we feel to improve our lives via the use of technology. We are now able to 3D print body parts like organs, retinas, and replacement hips. We are using the Internet of Things (IoT) to discover how the brain works, through sensors and charges to different parts of the body.[55] That information can now be used to help people born with paralysis. By simultaneously implementing an artificial neural network with

AI, robotics, and IoT with the human body, technology of the Fourth Industrial Revolution even helps people who were paralyzed from spinal injuries to walk again.[56]

Aren't these types of advances phenomenal? We must start debunking the myth that AI is somehow inherently evil and steer the conversation toward the positive aspects of the Fourth Industrial Revolution.

It's also important to keep in mind that just because the tenets and constructs mentioned above center on social causes, that doesn't mean that we must make a trade-off between social benefits and economic ones. These social causes are lucrative. They should be harmonized and reconciled in legislative conversations so investors can help bring these technologies forward. This is how we heal society of the wounds from which we are currently suffering.

THE GROWTH OF SMART INFRASTRUCTURE

The Moravec Paradox says that there are certain high order reasoning functions that machines do much faster. Take math, for instance. Yet when it comes to lower-level sensory skills like identifying or looking for things, people can do them much faster and more effectively. It sounds illogical to us, but for AI, high order reasoning is easier than low-level sensory skills.[57]

For example, if I were to put you in a room and ask you to identify the five exits in that room, it wouldn't take you very long. You can see the doors and the windows quite easily. This same job is much more difficult for a computer. To perform the same task, it needs to scan the room and then do the necessary calculations to identify which openings are windows and which are doors.

So computers can do math better, but humans are better at picking out exits. However, while a human will always need time and effort to do complex higher math functions, a computer can improve its performance, even at the sensory-motor tasks. With enough sensors, a machine could be just as quick as a human, and sensors are becoming more and more affordable. Even though there are many things that machines can't do right now, with just a few tweaks, they will be able to do them soon enough. As a result, technology will continue to grow and become more pervasive in our everyday lives.

This will undoubtedly be true in the case of smart infrastructure. The name refers to managing a city with AI. In most cases, AI sensors can collect data more quickly and holistically than traditional methods for analysis, and the information is then relayed into the cloud. City planners and infrastructure providers can access this data to use for maintenance assessment and urban planning. As an example, when it comes to city travel, the ability to accurately monitor just how long it takes to get from point A to point B helps governments make decisions on public transit options, such as when and where to add new bus routes.

Sometimes, human involvement isn't even needed. With sensor technologies integrated into a city's infrastructure and equipment, things like traffic, travel time, public safety, and utility usage can be monitored quickly and easily through technology alone. For instance, adaptive traffic lights that are able to assess the amount of traffic in real-time at a specific intersection or highway can adjust the timing of its lights in order to help the flow of traffic move faster. In the city of Bellevue, Washington, adaptive lights have decreased travel times on one main thoroughfare by 36 percent during peak rush hour. On another busy street, travel times have been reduced by 43 percent between the hours of 2 p.m. and 6 p.m. What could you do with that kind of extra time? According to city officials in Bellevue, the adaptive traffic light system has saved commuters 600–800 hours of time per year.[58]

Similarly, digital license plate readers placed throughout a city make it much faster to scan license plates for stolen cars and expired registrations. They can also alert law enforcement agencies if a license plate has been flagged for a warrant or violation. Beyond traffic control and safety, smart infrastructure extends into utilities.

When it comes to water usage, smart meters help manage resources and can identify leaks by comparing data collected over time on consumption and popular usage times. In short, a smart grid is a digitized version of the electric grid and runs communication between a utility company and its customers. The smart grid has sensors along transmission lines and responds automatically to changes in electric demand. Smart grids make energy more efficient and reliable. Operational costs are lowered for power utilities, ultimately reducing the cost of electricity to customers, and during outages and equipment failures, smart grids can automatically reroute energy to get electricity back quickly. AI technology in smart infrastructure improves our lives daily and is something to benefit from, not something to be feared.

For additional features made possible through smart grid infrastructure, see Figure 1.

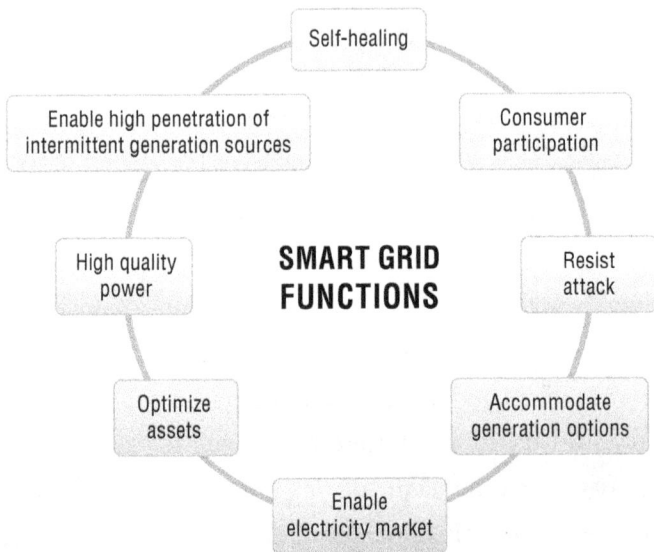

Figure 1. Smart grid functions

THE FIFTH INDUSTRIAL REVOLUTION AND BEYOND

The Fourth Industrial Revolution isn't simply a precondition for the fifth or sixth, because progression doesn't depend on this sequential numbering. Instead, economics operates on the logic of recognizing the unique historical moments where rapid change and convergence occurs. In that sense, naming this the Fourth Industrial Revolution is unnecessary. We could simply call this era the Convergent Technology Paradigm or even The New Paradigm. But there is some power to matching revolutions with numbers, as we've been doing it for centuries, so we continue to do it.

The first two revolutions involved what we call general purpose technologies (GPT): steam, electricity, combustion, and water plumbing. Over time, they stopped functioning as technologies and became the infrastructure of society. Then the Third Industrial Revolution brought us computing and the internet, which are quickly becoming GPT. Eventually, as the Fourth Industrial Revolution gets more traction, technologies like 3D printing, IoT, and AI will also become GPT, intractably embedded in our society.

The next revolution will certainly be called the Fifth Industrial Revolution, even if, like the fourth, it isn't a technical one. It will also be defined by what technology can achieve and how it keeps combining and meshing with humanity. The question is whether it will be a sequential or consequential leap from where we are now.

These revolutions become a fundamental aspect of society, even as they alter our values and purpose as humans. We can help ease this transition by preparing the financial industry to adopt new technologies. When we shift the conversation towards the financial industry, people start to recognize the emerging investment opportunities. It's also necessary to boost understanding that investment in new technologies doesn't really generate returns in a straight line. We need to push stakeholders to start expanding their horizons a bit wider so they can see the benefit of their investment on a grander scale.

Governments from around the world—industrialized and developing nations alike—who take on these initiatives will gain an economic advantage while serving as an example for other governments. At the same time, government leaders will need to rethink their role in society as technology changes the social order, including what jobs will mean in the future.

If we don't expand our ideas about future technological advances, we are limited to a very transactional use of technology, solely for increasing efficiency. We will be stuck with technology for the sake of technology rather than a redefinition involving social mandates, the nature of a job, and the design of jobs. We can and should do so much more.

PART II

HOW AI WORKS IN BUSINESS

CHAPTER 3

———

CURRENT AI TECHNOLOGIES

The best place to start, as they say, is the beginning. AI technology has gone through a lot in the past half of a decade. That's right; AI has been around for quite a long while. Yet its evolution and development have been a fascinating journey themselves.

In 1955, a young assistant professor called John McCarthy coined the phrase "artificial intelligence" and organized the famous Dartmouth Conference, which to many people marked the beginning of AI as a field of study. Since then, there have been two different methods to creating computer-based intelligence.[59]

RULE-BASED AI

The first is akin to how many adults approach learning a foreign language—we learn by memorizing the grammatical rules. The same concept applies here. The first AI machines operated based on a series of rules dictated by programmers. In effect, machines attempt to abstract the human expertise into decision trees built on explicit if-then rules. This will enable a machine to perform complex tasks like playing checkers or translating text. Indeed, this logic-based approach was initially so promising, it was thought it would dominate the development in the years to come.

Yet this first attempt in building AI turned out to be flawed. Let's use one of the most cited examples to see why. It works like this: the following sentence was translated into Russian, and the result was then translated back to English.[60]

The spirit is willing, but the flesh is weak. → The whiskey is strong, but the meat is rotten.

Despite the fact that the translated result is similar to the original and the words are correctly translated, the machine got the meaning completely wrong. This failed attempt revealed that a rule-based system for building AI had at least two major problems.

The first problem is that there are simply way too many

rules in the world, and knowing and being able to follow most of them is not enough. In fact, a system must get all of them right in order to perform well. Just like in learning a new language, getting only 80 percent of the grammar right will not do in many life situations.

A further obstacle is that there are quite a lot of sub-rules or rules-within-rules. In English, do two negatives make a positive in the phrase "don't tell nobody?" Would you confide the secret or not? If someone says, "I couldn't care less," does that mean she is considerate? How about two positives in the same sentence, would it connote a negative? No? "Yeah, right." The subtleties of these rules-within-rules make it impossible for these machines to achieve true intelligence.

No digital system can incorporate enough rules to fully understand how the world works. Indeed, this leads to the second, potentially insurmountable, problem, which relates to the so-called *Polanyi Paradox*, which can be summarized as: "we can know more than we can tell." Take learning to ride a bicycle, for instance. Here is the instruction on how to do so:

> When he starts falling to the right, he turns the handlebars to the right, so that the course of the bicycle is deflected along a curve towards the right. This results in a centrifugal force pushing the cyclist to the left and offsets the gravi-

tational force dragging him down to the right...A simple analysis shows that for a given angle of unbalance the curvature of each winding is inversely proportional to the square of the speed at which the cyclist is proceeding.[61]

It will be hard to find a more precise description. But does this tell anyone exactly how to ride a bicycle? Of course not, as there are many factors that ought to be have been taken into consideration but left out in the formulation of the rule. The tacit knowledge required to ride a bicycle isn't captured in words or rules and conveyed to others.

The same problem applies to AI: if many of the tasks we perform rely on tacit, intuitive knowledge, how can we codify the rules? In other words, if human beings themselves don't know the rules by which to accomplish something, then how would it be possible to create a rule-based system for any computer system to achieve the same accomplishments?[62] Consider getting a machine to identify a picture of a person or handwriting. If human beings themselves cannot explicitly articulate the features of the images, it will be hard, maybe impossible, to express them in executable codes for the machine.

While the rule-based AI got off to a promising start, these issues over such programming stalled the development of more-complex systems. This led to what is known as an AI winter, a period of reduced interest and little funding. By the 1980s, AI was mostly abandoned by developers because it was discounted as impossible to effectively achieve.

PROBABILISTIC VERSUS DETERMINISTIC

Clearly, if AI was to have a future, a different approach was needed. It is this very alternative that excels in the space of AI today, which represents a crude approximation to how human brains organically process information.

To understand how this works, consider again how we learn languages. As mentioned before, most adults learn a new language by memorizing all the rules, tenses, and conjugations. This can be very tedious and complicated, and it can be very arduous for adults to achieve true fluency in a foreign language.

Children, on the other hand, take a very different approach: they learn through immersive trial and error. Very often, they hear other people speak and emulate that. When they make a mistake, an adult gently corrects them. Children don't care about grammar rules; they don't even understand what they are. Gradually, they develop fluency, in a natural, fluid way, and usually at a level much higher than adult language learners.

While parallels can be drawn between the aforementioned rule-based, *deterministic* approaches to AI attempted and how an adult learns a language, current AI functions much more closely to how kids develop language skills. The AI learns by observing what it has seen in the past and makes the best effort to guess the right answer. When humans consistently tell machines *you're right* or *you're wrong*, the AI's accuracy improves significantly. In other words, over time the machines would gradually help themselves to improve the probability of choosing the correct answers.

For example, say we want to teach a machine to successfully identify pictures of dogs. Using the old approach, we would give it a list of rules: *dogs have tails, dogs stand on four legs,* etc. For this to work, we would need to flesh out all dog characteristics. But considering the great variety of dogs that exist, it would be hard to discern and define all the subtleties and exceptions that encompass a dog. This would be an exhaustive and ineffective exercise.

Using this approach, we would simply show the machine pictures, and it would begin to guess if the image is a dog or not. If it's right, we tell it so; if it's wrong, we tell it so. Based on those results, the machine changes the parameters it uses to maximize the likelihood of its predictions being true or minimize error and increase the odds of success in reaching the outcome. It's essentially a prediction machine[63] that has taken a probabilistic instead of a deterministic approach.[64]

In understanding AI this way, we can see why the "intelligence" is a misnomer. In view of the current politics around the world, we know for sure that there exists natural stupidity. But there is certainly no intelligence in "artificial intelligence."

LEARNING TO LEARN

In 1959, Arthur Samuel defined machine learning as the "field of study that gives computers the ability to learn without being explicitly programmed."[65] It involves computational methods that use experience to improve performance or to make accurate predictions. Experience, in this case, refers to past information or data that is available to us, which has been labeled and classified. As with any computational exercise, the quality and size of the data will be crucial to the accuracy of the predictions that will be made.

Looking from this vantage point, machine learning looks very much like statistical modeling. In statistical modeling, we first select a model and then collect data and go through the painstaking process of cleaning it (removing blanks and replacing, modifying, or expunging any incomplete, incorrect, or irrelevant parts of the data).

After that, we use this cleaned dataset as the input to test hypotheses and make predictions and forecasts.

In effect, we program the algorithm to perform certain functions based on the data we feed it. Put differently, the algorithm is static. It needs a programmer to tell it what to do when it is fed with data.

For machine learning, in some ways, the procedure is flipped. Rather than pre-selecting a model and then feeding data to it, in this case, it is the data that determines which analytic technique should be selected to best perform the task at hand. In other words, the computer uses the data that it has to select and consequently train the algorithm. Hence the algorithm is no longer static. As it is exposed to data, it analyzes the data, makes a decision, and then executes based on what it determines is the best course of action. In essence, it "learns" from the data, and in doing so, knowledge can be extracted from the data.

This method of learning is based on repetition (i.e., "training") using an algorithm. The term algorithm may sound complex. But it is nothing more than a set of instructions that is given to a computer to transform an input into a certain desired output. Thus, in machine learning, the learning aspect is just an algorithm repeating its execution operation repeatedly and making slight adjustments each time it does so until a certain set of conditions are met. The litmus test of a learning algorithm is when the final results will not change no matter how many times it is passed over the data. It is also important to note that there is no learning—the trained machines are merely trying to guess the right answers with increasing accuracy.

TYPES OF MACHINE LEARNING

Broadly speaking, there are three types of machine learning: supervised, unsupervised, and reinforcement learning.

The quickest way to understand the term *supervised learning* is that this algorithmic setup is *task-driven* in the sense that *you know what you are looking for*. As a result, it is possible to label the data to train the computer. Put differently, the data is already tagged with the correct label or the correct outcome. For example, if we were to teach a computer to identify dogs, then we would tag the picture of a dog as "dog" and non-dog picture as "no dog." This labeling process should be done by the programmer. Having learned the difference, the algorithm can now classify new information that is given to it and determine if the new image that it is seeing is a dog or not.

This approach can be used to read handwritten figures and alphabets. (Continental) Europeans and Americans write the number "1" differently, as shown in Figure 1. Indeed, even among Europeans, the number is often written in various ways. By feeding the computer with vast amounts of labeled examples of the "1s" or "As," it is possible to train the algorithm to see the variety of these figures. The computer begins to learn the variations and becomes increasingly competent at understanding these patterns. Today, computers are better than humans at recognizing such patterns of handwriting. The larger the dataset, the better trained the algorithm. Once trained, the algorithm is given new data and uses its experience to predict an outcome.

One last important note: even though computers can rather correctly identify pictures and characters, they have no concept of what a "dog," a "1," or a letter "A" is. Just as important, for the moment at least, is that while machines can do a good job at identifying what they are told to look for, they are unable to recognize what isn't labeled. Hence, the same machines cannot identify a "cat," a "2," and a "B." All AI can do is deliver the result of either "dog" or "no dog."

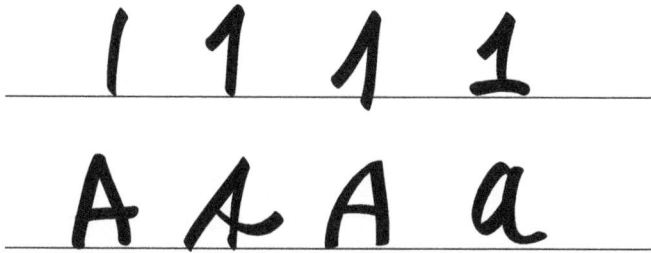

Figure 1. Variations of Writing "1" and "A"

A common application of supervised learning of today is training machines to identify credit card fraud. An example occurred to Terence over one Christmas. He woke up in the morning to find a message from his credit card company informing him that there was a transaction of £200 incurred the day before through a payment gateway, which he had never heard of before. A machine identified the charge as a potential fraud and flagged it for closer inspection. This kind of system makes up a big chunk of AI applications today.

By contrast, *unsupervised learning* is a data-driven method in which you don't know what you are looking for and leave it to the machine to see if it can figure something out. It is often used to make connections of which the human programmers were totally unaware.

In this approach, the algorithm is trained using a dataset that does not have any labels. The algorithm is not told what the data represents. In this case, the machine uses methods of estimation based on inferential statistics to discover patterns, relationships, and correlations with the raw, unlabeled dataset. As patterns are identified, the algorithm uses statistics to identify boundaries within the dataset. Data with similar patterns are then grouped together, creating subsets of data. As the classification process continues, the algorithm begins to understand the dataset it is analyzing, allowing it to predict the categorization of future data.

Unsupervised learning can be used to identify unusual internet traffic patterns or new forms of insurance fraud. This clustering of data can automate decision-making, adding a layer of sophistication to unsupervised learning. More importantly, it allows us to leverage data in a new way. What we lack in knowledge, we make up for in data. We may not know what we are even looking for, but as long as we have the data, we can find patterns and pull useful information.

A third but less common way for machines to "learn" is *reinforcement learning*. In this case, the machine is given "rewards" and "punishments" telling it whether what it did was right or not. The algorithm is presented with data that lacks labels but is given as an example with a positive or negative result. This positive or negative grade provides a feedback loop to the algorithm allowing it to determine if the solution it is providing is solving a problem or not. Effectively, it is the computerized version of human trial and error learning. Too abstract? We see it in real life all the time:

Imagine a baby is given a TV remote control at your home (environment). In simple terms, the baby (agent) will first observe and construct his/her own representation of the environment (state). Then the curious baby will take certain actions, like hitting the remote control (action), and observe how the TV would respond (next state). As a non-responding TV is dull, the baby dislikes it (receiving a negative reward [or "punishment"]) and will take less actions that will lead to such a result (updating the policy) and vice versa. The baby will repeat the process until he/she finds a policy (what to do under different circumstances) that he/she is happy with (maximizing the total [discounted] rewards).[66]

Perhaps the most famous experiment conducted on this type of machine learning was the engineers at Google's DeepMind, which trained the algorithm to learn how to play the video game Atari Breakout. At first, the machine was playing very clumsily. However, four hours into the game, it discovered that the most effective way to win the game is by digging a tunnel through the wall.[67] Indeed, it was reported that the machine had managed to discover at least one move previously unknown to the programmers.[68] All of these were achieved without the machine understanding what a "ball" or even a "game" is.

Application-wise, the key is that decisions lead to consequences; the output action is prescriptive and not just descriptive, as in unsupervised learning. Even though reinforcement learning is still in its infancy, it is believed to have huge benefit potentials in various situations such as managing traffic lights and offering personalized recommendations.[69]

NEURAL NETWORKS

Within the machine learning fields, there is an area often referred to as brain-inspired computation. Today's AI technology is often associated with (artificial) neural networks. The reason is that this technology is a crude emulation of our own human brains, our biological neural network, which is shown on Figure 2. As displayed in the figure, a neuron is made up of a cell body, dendrites, and axons. Information comes in through the dendrites; the cell body processes the information and passes it through the axon to distribute through the synaptic terminals. There is a very clear input (through the dendrite) and output (through the axon).

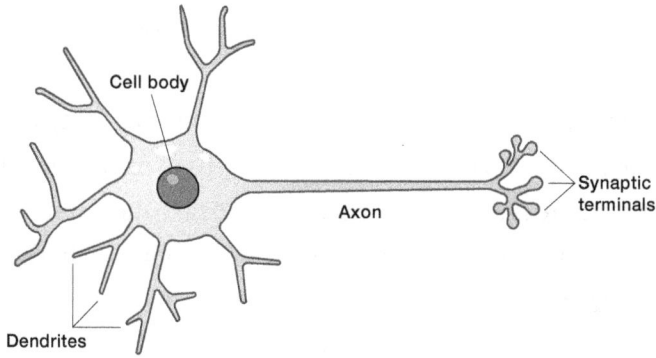

Figure 2. A Single Neuron

In Figure 3, modern AI technology is loosely modeled after the neuronal structure of the brain's cerebral cortex but on smaller scales, where there are inputs, output, and a processing body.

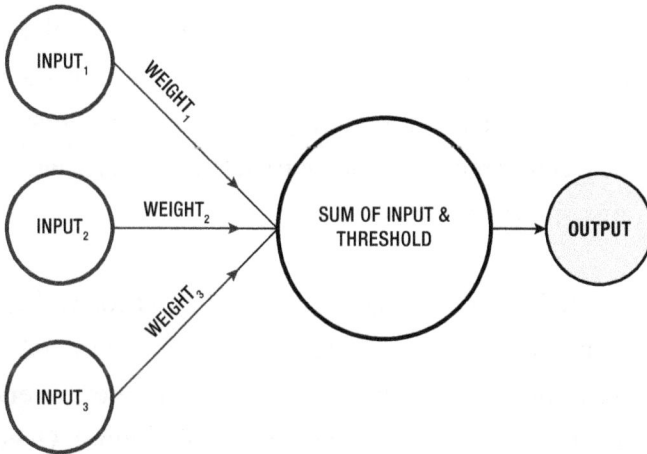

Figure 3. Node in Artificial Neural Networks

Let's quickly examine how we as human beings make decisions. Consider the following hypothetical: you are about to leave the house. You are pondering the question, *do I bring an umbrella with me or not?* You'd probably look out the window to see what the weather is like. If it's sunny, you'd determine there's no chance of rain, and you'd leave the umbrella at home. If it's overcast, you'd think there is a chance of rain, so you'd take the umbrella. The input was the weather, and the output was whether you brought your umbrella with you.

Of course, weather condition is never just sunny or overcast. If there have been several days of occasional clouds and no rain at all, you might abandon the idea of taking an umbrella based on the data you've "collected" over the last few days. Consequently, if you think that an input is irrelevant (such as a cloudy sky with sunny breaks), you mentally assign less weight to that input. By contrast, you may, based on your experience, attach more importance to something else (e.g., the trustworthiness of the weather report) or location specificities (e.g., when it rains, it rains hard).

However, if you don't take your umbrella and you end up stuck in the rain, then the next time you need to make the same decision, you may adjust and place more attention to the "cloudy sky with sunny breaks" input. Over time, you would continue to refine your approach by redistributing the weight of different inputs. Some you might discard entirely, whereas others might become more important.

Figure 3 illustrates how an artificial neural network is modeled after the human brain. Inputs 1, 2, and 3 are the information coming in. When an input goes into the cell, it will process it and push out the output.

In the context of AI, the machine will weigh the three inputs based on what it has "learned" and "seen" in the past. If an input is deemed to be unhelpful in reaching the successful outcome, then it is given less weight. If an input is consistently relevant, it is assigned more weight. If this process continuously iterates weight with more and more data being fed into the machine—or more simply, "trained"—the weight of the information that creates no value will be reduced to zero, eventually, while that attached to the important information will be pushed up. Consequently, over time, the AI machine will become ever better at guessing the right answer. Therefore, if a computer is looking at pictures to determine if something is a dog or not, it might weigh the presence of eyes as an important factor with heavier weight and the background as something that it doesn't need to factor in at all. With newly ingested information, machines can adjust the weighting next time, thereby increasing the accuracy of their responses over time—just as our brains do.

DEEP LEARNING

The cases in which to make decisions insofar have been

simplistic. But most situations in real life are much more complex than identifying a picture of a dog or deciding whether to grab an umbrella. If you're starting a company and you need to decide on a business strategy, for example, reaching the right decision would be much more complicated. To do so, you would compile and review a great deal of information, and you would then contemplate and think "deeper and deeper" about the issue. In effect, we are processing information over and over again across multiple brain cells until we are able to make a decision.

Current machine learning technologies work in the same way. If you are using a computer to figure out a complicated problem, you will want the computer to go through more processing. This is what is known as *deep learning*, as shown in Figure 4.

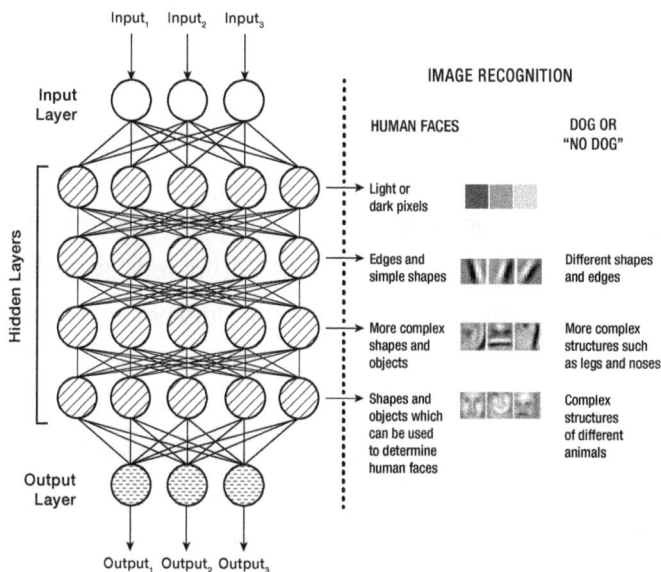

Figure 4. Deep Learning

In such a network, there is an input layer and an output layer. Everything in between is a hidden layer of artificial neurons, or *nodes*. Each layer processes information passed on from the previous layers of nodes and transfers the results to the next. Hence, the more layers in a system, the more processing will take place—the "deeper" the learning there will be. Deep learning therefore performs better at conducting the analysis and reaching a more accurate result. In that sense, deep learning is not about developing a deeper understanding but about processing data much more intensely.

The right-hand side of the figure illustrates how deep learning neural networks use layers of increasingly complex rules to recognize complicated objects such as faces. As an over-simplified example, the first layer can be thought of as the machine's trying to figure out pixels of light and dark. With the further layers, the computer identifies simple shapes and edges and subsequently more complex shapes and objects. Ultimately, by the last layers, the computer knows the shape and object that can be used for discerning faces of people.

The more layers there are in a network, the subtler the features of the input data that can be detected and identified. In its 2016 special report on AI, *The Economist* stated that networks twenty or thirty layers deep are common, and researchers at Microsoft have built one with as many as 152 layers. Deeper networks are capable of higher levels of abstraction and produce better results, and these networks have proved to be good at solving a very wide range of problems.[70] Figure 5 provides a summary of the concepts among AI, machine learning, and deep learning.

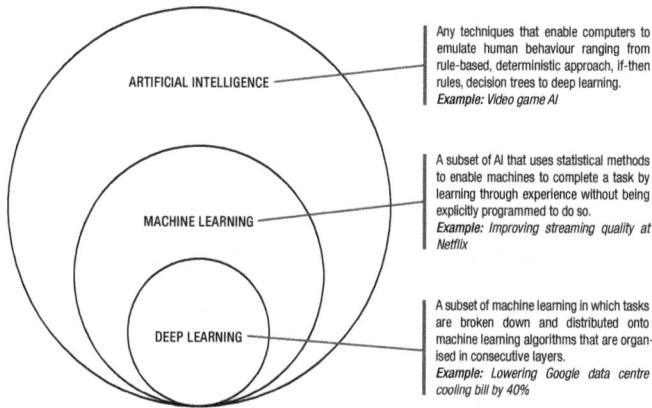

Artificial Intelligence: Any techniques that enable computers to emulate human behaviour ranging from rule-based, deterministic approach, if-then rules, decision trees to deep learning. *Example: Video game AI*

Machine Learning: A subset of AI that uses statistical methods to enable machines to complete a task by learning through experience without being explicitly programmed to do so. *Example: Improving streaming quality at Netflix*

Deep Learning: A subset of machine learning in which tasks are broken down and distributed onto machine learning algorithms that are organised in consecutive layers. *Example: Lowering Google data centre cooling bill by 40%*

Figure 5. Artificial Intelligence, Machine Learning and Deep Learning

BRINGING IT CLOSER

The technical applications of AI have only recently become a part of the public consciousness as machine and deep learning technologies have begun to play an increasing role in our modern lives. If you go back more than a few years, most people had only heard of AI from a pop culture perspective, in movies like *The Terminator* or 2001's *A.I. Artificial Intelligence* directed by Steven Spielberg.

So what really brought about the current developments of AI? How did it go from relatively little to suddenly everywhere? After all, the origin of neural networks described above is based on a mathematical model of the neuron called the Perceptron, introduced by Frank Rosenblatt as early as 1957. Rapid developments in two areas have contributed to the latest rise of AI: processing power and data.

Speedy improvements of computer chip technologies have led to a great increase in computing power, which, in turn, allows for expansion of processing and analytical capabilities. Yet the real boost came from another parallel industry development. The introduction of the graphic processing unit (GPU) made popular by the rise of PCs and video gaming consoles dramatically sped up processing by nearly a factor of a hundred. In one case, it was reported that training a four-layer network using GPU now took less than a day, as opposed to several weeks in the past.[71]

The recent phenomenon of Big Data and its availability has created a lot of data for training machines. Consider graphic images, for instance. A study in 2014 suggested that people uploaded an average of 1.8 billion digital images onto Facebook, Instagram, Flickr, Snapchat, and WhatsApp *every* day.[72] That equates to a whopping 657 billion photos a year. Put differently, every two minutes, humans take more photos than even existed in total 150 years ago.[73] These, of course, are not including the digital text, videos, sensor readings, voice, and sound recordings that can also help train machines. Just like a child learning a new language needs a lot of words and sentences, machine learning systems now have a lot of examples to enhance their ability to recognize speeches, classify images, and make sense of text.[74]

Storage of data and code has also made a lot of progress. The arrival of cloud computing allows for a massive amount of data to be stored for future use. Cloud computing, such as that offered by Amazon Web Services, has created new opportunities to develop AI, particular for those startups with smaller budgets.[75]

BAD BOTS?

While there are many reasons for us to congratulate ourselves for the technological triumph humans have achieved insofar, it is paramount to highlight the downside. In the past, machine learning has led to unexpected or unwanted outcomes, particularly when the input the machine has received is (perhaps unintentionally) incomplete, poorly chosen, obsolete, or subject to selection bias.[76]

In one recurring example, programmers creating facial recognition software thoughtlessly used photos solely of Caucasian people to train the AI.[77] As a result, in 2016 in New Zealand, a man of Asian descent had his passport photograph rejected when facial recognition software mistakenly registered his eyes as being closed.[78] In another case, in 2015, an Afro-American software developer tweeted that Google's photos service had labeled photos of him and his friend as "gorillas."[79] For certain, the machine wasn't inherently racist, but it was fed biased data, and the outcome reflected that.

Flickr did something similar when its new auto-tagging system labeled Afro-Americans as "ape" and "animal" as well as pictures of concentration camps as "sport" or "jungle gym."[80] AI, as it turns out, can also be sexist. In one study, people featured in kitchens are more likely to be labeled "woman" than reflected the training data. In fact, in one case, a photo of a man at a stove was labeled "woman."[81]

More unintended reflections of the worst side of humanity occurred when Microsoft released on March 23, 2016, a public chatbot with a Twitter account, named Tay. This was a stunt to show off the chatbot's autonomous abilities when left completely to its own devices. Tay was supposed to learn based on the input it received from interacting with other internet users. Within hours, Tay was flooded with abuse from other users. Microsoft had to pull Tay off the internet after just sixteen hours, because it had begun spouting verbally abusive, racist, sexist, and pro-Nazi sentiments. Tay was poisoned by ugly data input (which says a great deal about some social media users), and Microsoft had not programmed the bot with an understanding of appropriate behavior. Quite frankly, this incident says as much about us as human beings as it does about machines.

These examples show how important it is to program AI in a purposeful and conscientious way. The bots aren't bad; they've just received bad inputs. If we want to use these machines to create benefits, we must also be aware of our own and other people's biases when programming. As long as AI systems learn from flawed human data, we can expect algorithms to automate bias, discrimination,[82] and inequality.[83]

Blind application of machine learning increases the risk of amplifying biases in data. Yet very often companies are more eager to push out new AI-driven services than deal with the details. Three years after Google's photos service issue mentioned above, it appears that the company "fixed" the problem by merely removing certain labels from the system, including "gorillas" and "chimpanzees."[84]

This is not an issue to be taken lightly. Let's draw an analogy: the fact that many people are laying their trust on feeding the algorithm to decide on our behalf is akin to giving data and information to a six-year-old child with the expectation that she will be able to process it and come up with an output to make decisions for the adults' world. The child lacks the necessary sophistication in her analysis to reach results that can meet the complexity of logic underpinning and the reality of adults' activities. Furthermore, the child is incapable of assessing the quality of the supplied information and sees the world only through the perspectives prescribed by the adults. If the adults cannot pick out flaws in their views, it is inevitable that the child—or algorithm—will perpetuate the flaws. Just like children are not naturally born holding prejudice; they learn it first from parents and continue to deepen their views through a widening circle of people and institutions ranging from extended family members to schools.

The issue around bias is not only pertinent to the private sector, as an increasing number of government agencies are starting to deploy some resources to the study of bias and biased outputs. Worthy of a mention is the initiative called AI for Humanity, introduced by the French government with the ambition to set an ethical mission and a mandate around AI strategies. Concepts like transparency in the programming and loyalty around the algorithms are some of the interesting elements visible in this effort. While still at its early stages, it is a welcome development that we wish to see proliferate in other parts of the world as well.

AI AND THE INTERNET OF THINGS

The term "Internet of Things," or IoT, was originally coined by Kevin Ashton in a presentation in 1999. It refers to the interconnection of computing devices embedded in everyday objects. Nowadays everything from your cell phone to your washing machine can send and receive data over the internet. This is raw data that can be utilized for data analytics and possible interpretations of AI, but the most important aspect of IoT is the fact that billions of sensors around the world are getting connected to each other and to the cloud.

These very same sensors have been used since the beginning of this century to improve performance and efficiency in the value chain, but they have also generated access to measurements of data flows and detections of errors or production flaws, as well as specific behaviors that can help improve performance outputs. Here's a simple example: imagine a fridge that can measure the availability of a specific product—say milk. If we run out of milk and do not replace it with a new bottle, a "smart fridge" could send an email to the grocery store placing the order for a new bottle of milk, paying for the order with a preregistered credit card, and providing the delivery address. This is a trivial example, but it exemplifies the degree of connectedness of IoT and its potential.

The integration of all this data on one side and the practically infinite virtual storage is a key feature of what IoT and its applications can do.

While the example above is still confined to a plausible but equally hypothetical scenario, the place where IoT has been mostly utilized has been on factory floors to improve efficiency of production. When IoT is shifted to its application within an industrial context, it is referred to as Industrial Internet of Things or IIoT. Companies like UPS, DHL, and Amazon employ automation, sensors, and driverless transportation in warehouses to track and move parcels. They connect physical objects to the virtual world using sensors and run their businesses this way. If you want an example from a company in the UK where IIoT has become the daily bread and butter of the production processes, you just need to visit the warehouse for the online supermarket Ocado, which uses a lot of sensors to automate otherwise very labor intensive processes.[85]

AI and the IoT are not the same thing. Whereas the former is more about data processing and analysis to predict results, the latter relates to connectivity between devices. Being able to work at the forefront of AI, we are privileged enough to have front row seats to seeing and experiencing firsthand the latest technological developments in the field. To us, the milestone achievement in AI is not just the fact that we can now collect even more data but also the newly gained ability to convert unstructured data into purposeful structured data. In our everyday lives, we are constantly generating data, but most of it is unstructured, uncategorized, and hence generally unusable. One thing AI has done is allow us to extract huge amounts of data sources and situations previously difficult, if not impossible, to compile, and process and make sense of them.

Imagine what and how much can be accomplished at once if you can have an instant count of the number of people there are in a specific defined space? Or automatically detect when an accident (e.g., someone fell) or incident (e.g., a fight) occurs? A partner company of ours called DT42 is using AI for object and event recognition, as shown in Figure 6.

AI FOR OBJECT & EVENT DETECTION IN PUBLIC AREA

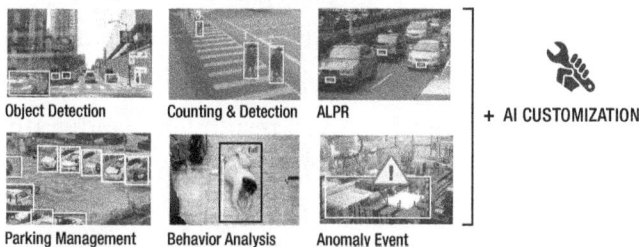

| Object Detection | Counting & Detection | ALPR | + AI CUSTOMIZATION |

| Parking Management | Behavior Analysis | Anomaly Event |

Figure 6. Potential Use of Edge AI

For example, it uses cameras to observe how many pedestrians are crossing a street versus how many cars are turning left at an intersection to improve road safety. In the past, the data was already being generated by pedestrians and motor vehicles, but only recently it has become possible to capture it, analyze it, and draw useful conclusions in a timely fashion to prevent accidents.

A solution to this problem is to use machines to process

the data and take immediate response actions. Yet the fact that data is first captured by cameras and then sent for processing by machine learning systems in the cloud suffers from two problems. The first is bandwidth. It is expensive because the bandwidth to handle the communication of massive amounts of data between cameras and the cloud is always limited. The second is latency. The time required to move data to the cloud, process it, and then move it back to devices is often far too long for instant response.

This is why *edge computing* has garnered so much attention lately. This is an expertise of DT42. Calling it edge AI, the company places the AI processing capability in the cameras themselves. As a result, less bandwidth is required as less data needs to be sent to the cloud. On the other hand, latency is also reduced since the time it takes to transmit data back and forth can be effectively eliminated.

With the expanded use of IoT, we can expect to see more and new AI technologies be deployed. Yet as much as people like to talk about the IoT, it still has a lot of technical and operational details that need to be worked out. Establishing the use of IoT is, contrary to what many people believe, not really that simple. This is because for sensors to work well, two critical elements are needed: power and connectivity. To most consumers like us, they are widely available. But there are many situations where a desired sensor is either too far from a power source (such as in a tunnel) or infeasible to place (such as on the actual ground support equipment at airports). Of course, they can be powered by batteries. But this would mean you will have to go around to check and replace them on an ongoing basis. Additionally, 3G or 4G connectivity is extremely expensive for the purpose of connecting sensors, not to mention such connectivity needs to be powered itself.

A company called EverSensors has come up with a product called "EverSense," which requires very little ambient light from the surrounding to work, thereby dispensing with the need for batteries and power sources. The company's slogan is "Stick and Forget. Works Forever." On the connectivity side, these sensors don't really need to stay online on a consistent basis or upload large chunks of data. As long as the sensor can collect enough power from the immediate surroundings to do one single data burst.

All of this sounds mundane, but the sensor can be used to perform previously impossible tasks—for instance, temperature monitoring of freezers in convenience stores. Some ice cream manufacturers in Hong Kong, for instance, typically own the freezers in stores and would want to keep track remotely to ensure they are in proper working order so as not to damage the ice cream products contained therein. Yet it would have been difficult previously because batteries don't really work in cold conditions, and therefore electricity-powered sensors don't function well in such settings. EverSense is perfect for this occasion and opens up new possibilities and business opportunities.

With the refinement of sensors and cheap connectivity technology together with AI, we will be able to better manage things like infrastructure and physical assets. This will translate into longer life cycles of assets, which saves a lot of money and unnecessary effort. Consequently, we may end up disposing of less, which is great for sustainability and the environment.

The combination of mobile connectivity, Big Data, the cloud, and machine learning has been a powerfully successful technological advancement that has rapidly penetrated our world. AI has come a long way but still has a long way to go to be employed worldwide. Undoubtedly, AI will continue to change our lives and have vast business and societal implications as well. This is what we will explore in the next few chapters.

CHAPTER 4

AI IN BUSINESS

Despite the current limitations of AI as described in the prior chapter, the technology is bringing broad changes to the business landscape. We are starting to see AI employed more and more in the context of business activities. Revenue derived from AI-related product and service offerings is expected to exceed US$3 trillion by 2024, up from just US$126 billion in 2015.[86] A recent study has found 85 percent of 3,000-plus executives surveyed believe AI will allow their businesses to obtain or sustain competitive advantage.[87]

In many ways, AI *seems* like the perfect new technology for business integration. It has many potential business applications because of the inherent sense of efficiency it creates. This boost in efficiency is very attractive because it often leads to reducing operating costs.

Previously, many businesses attempted to boost efficiency by trying to reduce mistakes. The Japanese business mindset introduced a culture of continuous improvement and minimizing errors as close to zero as possible during production. This same philosophy of reducing mistakes also spawned other management techniques like Totally Quality Management (TQM) and Six Sigma. Companies that employ AI must take a different tack.

AI is terrific at preventing errors, especially in labor-intensive business activities such as manual data entry or any kind of repetitive tasks, and the money that can be saved on this alone is enough to justify employing the technology in this capacity. To that extent, AI is a perfect economic trade-off: more efficiency that over a period of time will recover the cost and generate more profits. But there are many other ways that AI can help improve a company's effectiveness.

For instance, it can help companies access new revenue opportunities such as chatbots and better recommendation services. Just as important, companies will increasingly find new ways to leverage AI to create profitable new business models. This is because machine learning is seen to be a "general purpose technology" or GPT—technologies that drive most economic growth by unleashing cascades of complementary innovations, like internal combustion engines,[88] and that have the cunning ability to quickly trigger economies of scale. Andrew Ng, one of the most renowned experts in the field, likens AI to another GPT: "Just as electricity transformed almost everything 100 years ago, today I actually have a hard time thinking of an industry that I don't think AI will transform in the next several years."[89]

There is currently a gap between the implementation of full-fledged AI solutions within larger organizations versus within smaller organizations. For reasons we will discuss later in this chapter, smaller organizations usually can't access these innovations, due to high cost and lack of expertise. However, they should be able overcome these issues very soon. This, in turn, means that many businesses, small or large, will have the chance to put AI into their operations.

This chapter covers how AI can work within a business, while the next looks at how to apply it in your own.

"FITTINGNESS," NOT JUST FITNESS

The key to reaping the most benefits from using new technologies does not lie in the fitness or prowess of the idea and technology. Rather it is the "fittingness"—how well they are integrated into the existing activities that lead to new value creation. For AI, integrating the technology into your company does not simply mean replacing human tasks with those performed by machines. It is infinitely more effective to integrate AI into your business operations while still employing humans to do the tasks they excel at.

Amazon offers a terrific example of how to use machines to increase efficiency, while still relying on humans to perform certain key tasks. One might think that the entire process of an Amazon Fulfillment Center, where customer orders are assembled, would be fully automated. But it's not. Terence was lucky enough to have visited one of the largest among them and was pleasantly surprised to discover that many tasks are still undertaken by humans.

When the products first arrive in the warehouse, they must be shelved. As it turns out, humans are better at shelving than robots. If you have a stack of books, a machine will try to fit them in the most efficient way onto the shelf, without regard to what direction they are facing. That's not how we want books to be shelved, however; we need them to all be facing one direction, with their spines facing outward for easy visibility. The same is true when you're stacking objects of different shapes and sizes; you want everything to be easily viewable while still conserving space as much as possible. You want books to be standing with each other and not just stacked one on top of another to maximize storage space. Amazon tried to automate this process, but AI doesn't understand space like this. The amount of size and shape variation between objects was just beyond the ability of robots. Dealing with irregular forms of products is something that humans excel much better at.

Now let's look at the products going out. When a customer order comes up, one would imagine a tech giant like Amazon would have a digitally driven mechanical arm to grab products of all shapes, sizes, and consistencies and put them in a box for shipping. Yet the company has not yet found the right way to do so even after various attempts.[90]

When pulling items to ship, humans again outperform machines. Instead of human staff running up and down the fulfillment centers to get to the shelves to pick out the right items, Amazon has instead used some knee-high, orange-colored robots called pods that bring the shelves to the human pickers. Once the shelf is brought in front of the picker, she then picks the ordered item from the shelf. Next, she checks the picked item against the image of the order on the screen to confirm the product matches the shipping order. After that, she drops it off in one of the five bins (with each dedicated to one of five customer orders being filled at the same time), signaled by a light above the bin to indicate in which bin to place the item. She then presses the light, and the robot scoots away with the shelf. The next robot carrying another shelf will then take up the position in front of the picker.

There are several good reasons for this arrangement. First, humans are much better and quicker than machines when it comes to checking product match. Machines will have to scan the item and analyze it, which slows down and adds cost to the fulfillment process. Second, human hands are more dexterous in picking off products with delicacy. Third, humans are much better at quality control: they can easily see if a book is ripped, folded, or dog-eared or if the wrong title is in the designated space. Returns due to damages or a wrong order are expensive for the company, and reducing return orders is paramount to the profitability of the business.

A fourth reason has much less to do with humans versus machines. Yet it is genius enough that we must list it. As a result of being able to bring shelves to humans, copies of the same item can now be scattered around and stored anywhere in the warehouse as long as the company knows which shelves they are on, as opposed to packing them all in a single location, which allows for flexibility in using space.

The story does not end here, however. Surprisingly, there is also an army of human staff to pack the shipment boxes. Once again, one would imagine Amazon uses machines to do such tasks. And once again, the reasons they do not are these: the dexterity offered by humans in packing orders of different sizes is unmatched, and further quality control checks. The more human touchpoints there are, the better, as the return rate will be lower.

With these arrangements, Amazon is taking full advantage of what human beings can offer and leaving the literal heavy-lifting acts of moving up and down the warehouse aisles to machines, preventing physical exertion and time loss that human staff would have incurred in performing the same tasks.

This is a great system because it really capitalizes on the strengths of humans *and* those of robots. There are limits on what robots can do, although that might change someday. Until then, humans are the best candidates to fill those gaps given our innate flexibility. Humans can increase the process's efficiency by reducing errors. Contrary to what you might think, Amazon tries to maximize the amount of times a human being touches an order for this reason. This is a great example of cobotics, robots intended to exist alongside and physically interact with humans in a shared workspace, something we mentioned earlier in the chapter. A sort of noninvasive professional symbiosis between humans and automation.

The illustration above concerns humans and machines

working together to achieve better outcomes in a physical setting. How about people working with AI? In a recent study, professionals used deep learning to identify cancerous biopsies on images. Machines could attain an accuracy of 92.5 percent. But humans had an accuracy rate of 96.5 percent, so machines still have a higher margin of error. The truly interesting thing is when AI and pathologists worked together, the margin of error went down to 0.5 percent, effectively boosting the cancer screens to nearly 100 percent accuracy.[91]

Once again, this goes to show that having large amounts of data or machines alone doesn't necessarily mean better results. The best results occur when a human is using an algorithm to inform their decisions rather than an algorithm (or a human) making the decisions on its own.

THE TECHNICAL FUNCTIONS OF AI

One important observation from the discussions so far is that technology can excel in certain areas while humans are better at performing others. So it's important to understand the strengths of AI so that you can properly utilize it for the correct tasks. To do so, as shown in Figure 1, it is possible to dissect AI and look at its three distinctive technical functions separately: recognizing, predicting, and prescribing.[92]

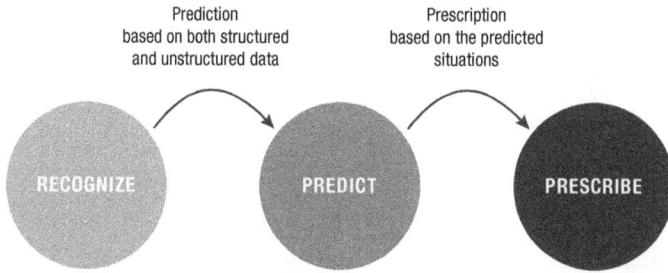

Prediction
based on both structured
and unstructured data

Prescription
based on the predicted
situations

RECOGNIZE   PREDICT   PRESCRIBE

Figure 1. Three Technical Functions of AI

RECOGNITION

As detailed earlier, using machine and deep learning systems, it is now possible to extract data that was previously tremendously difficult to collect. Gradually, we can now use AI to compile data that presents itself in the form of images, videos, voice, and text.

Such data has always been in existence and generated every day. Yet such idle and unstructured data has never been until now easily captured. On top of it, processing of such data has improved at neck-breaking pace over the past years. Take image recognition, for instance. Figure 2 provides a snapshot as to how the margin of error in using deep learning to recognize images has come down quickly.[93]

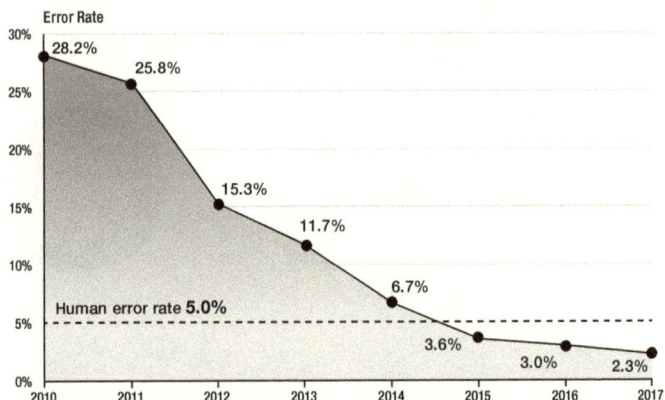

Figure 2. The ImageNet Large Scale Visual Recognition Challenge Results 2010–2017

In 2010, when the ImageNet Large Scale Visual Recognition competition started for the first time, every team got at least 25 percent of their results wrong. The first breakthrough came in 2012, when the team to first use deep learning ended up being the only team to get its error rate below 25 percent. By the following year, nearly every team made it to having a margin of error of 25 percent or less. The second breakthrough came in 2015, when machines outperformed the 5 percent error rate of human beings. In 2017, twenty-nine of thirty-eight competing teams got less than 5 percent wrong.

Speech recognition at the same time has also continued to improve. In 2013, the voice accuracy rate of Google machine learning was somewhere between 70 and 80 percent. By 2015, it had reached 90 percent. In 2017, the machine was on par with human beings at 95 percent accuracy.[94]

Such advancement has also naturally given a great boost in the development of *natural language processing*, more popularly known as NLP. The term refers to the ability of a machine to analyze, understand, and generate human speech and text. The ideal goal of NLP is to make interactions between machines and humans feel exactly like interactions between two humans. Again, capturing speech and text is relatively easy; the hardest part is to understand it—to make sense of them. For instance, figuring out semantic and contextual meanings of conversations is not easy for machines. Yet recent development of machine learning models such as "word2vec" have broken new ground in this area.

PREDICTION

The second function is prediction. As it was made clear in the previous chapter, current machine and deep learning systems run on probability, continuously improving themselves to make the right prediction based on processing both the structured and unstructured data. Prediction can therefore be thought of as "the process of filling in missing information and using it to generate information you don't have."[95] Most businesses still run on conventional mechanisms: they do their forecasting and predicting without the aid of technology. At the current pace of advancement, we will likely see more and more financial industries automate their processes to improve relationships between banks and account holders, such as adding more rigorous measures to prevent credit card fraud and anticipating customer's cash needs.

When Augie Picado, President of UPS Mexico, addressed Mark's class at Harvard, he shared some of the AI applications to trade companies like UPS, DHL, and FedEx. It was fascinating to find out they use AI to determine whether packages contain any contraband that should not be shipped. These companies process millions of packages a year, and there is no way they could manually check each one, so they use technology to scan all parcels for counterfeits goods, live animals, drugs, and so on. UPS even has robots that roam its warehouses, looking for counterfeit goods, just like drug-sniffing dogs.

Another example is the use of AI to monitor buildings and machines. The so-called predictive maintenance— forecasting when and what to repair before the breakdowns—would likely lead to massive improvements in the use of infrastructure and many tangible assets.

One of the most important values of prediction is that it is a foundational input. Better predictions, in both our business and social lives, can lead to better information and, in turn, better decision-making. As machine learning systems become more and more popular, the cost of prediction will continue to fall. A result is that we will be coming up with a diverse range of new activities in addition to enabling all sorts of things that were previously impossible to do.[96]

PRESCRIPTION

But prediction is not decision—it is only a way to inform decision-making. How the predicted outcomes are used is the third technical function—prescription. Drawing on a lot of sensors, connectivity, and machine learning systems, AI has been applied in autonomous vehicles. The same can be said for advanced robotics, potentially supercharging the automated industrial production. Knowledge-based office activities will also see intelligent automation, which is about enhancing workers, both human and digital, by embracing and working alongside intelligent technologies.

One important element to consider with prescription is how much AI will be replacing human judgments. Machines are ideal for issues that are of little consequence, such as deciding to disapprove a credit card transaction. A mistake would be an annoyance, no doubt, but certainly not life-changing.

But imagine a situation in which you have a choice between humans or machines passing a legal judgment on you. Humans are often known to be flawed. A past

study has shown that chances of receiving paroles gradually come down as the judges get closer to their lunch
break, to name an example.[97] At the same time, AI is
subject to its own programming biases, some of which
are not immediately evident. Knowing this, would you
choose machines or humans to make the judgment and
come up with the verdict on you? Decisions on how to
make proper judgments in the AI world will become even
more of an issue to confront.

AI WAVES

AI has allowed us to unleash our imagination as the technology has entered so many different aspects of running
businesses and our social lives. To enable us to better pin-
point the various categories of AI, we find the four waves
of AI, proposed by Kai-Fu Li, a well-known expert in the
field, to be very handy. They are shown in Figure 3.[98]

AI WAVES	CATEGORIZATIONS	USA	CHINA
1 Internet AI	Apply AI to serve consumers through web and mobile services with enhancement of user interface and of understanding of consumer preferences	*Lagging*	*Ahead*
2 Business AI	Apply AI to help businesses make better decisions, understand customers, marketing, or reduce costs and workloads with intelligent automation	*Ahead*	*Lagging*
3 Recognition AI	AI for voice / image / video recognition to extract information that was never captured before and convert into usable data for creating new systems and applications	*Lagging*	*Ahead*
4 Autonomous AI	AI to automatically move initiatively with the ability to sense and respond to the surroundings such as autonomous vehicles, robotics and manufacturing, and production automation	*?*	*?*

Figure 3. Categorization of AI

The first wave is "Internet AI," which began just before the millennium, with the introduction of algorithms to learn from the ever-accumulating user data to tailor personalized content and preference engines. This involves a great deal of learning and achieving a better understanding of consumers through continuous improvements in user interface and apps. Companies can capitalize on the masses of labeled data to put together pictures of personalities, habits, demands, and desires of their customers.

The second wave, starting circa 2004, is "Business AI." Whereas in Internet AI algorithms are used for understanding consumers, Business AI is more about business process reengineering. It makes use of labeled data that companies have been accumulating. Transactions for financial services and machine maintenance records for industry, for instance, all generate data to inform companies to make better decisions. While humans can make predictions based on clear root causes (also called strong features), the arrival of machine learning is now allowing the detection of correlation among variables not readily identifiable by humans (also called weak features). The result is that new perspectives are opening up that can complement human judgment.

The third AI wave, "Perception AI," emerged around 2011. We prefer to call it "Recognition AI," as we believe that term to be a more precise description for what it is.[99] This involves digitalizing the physical world using sensors and smart devices, compiling data that was previously hard, if not impossible, to capture before. As discussed above, the ability to make sense of new sets of data led to many applications and business opportunities.

The fourth wave, which began around 2015, sees AI being used in the physical setting instead of mere software-driven digitalization of the world, effectively bringing all three previous waves together. Here, AI equips machines with the abilities to sense and react to the world around them, to move intuitively, and to handle objects as easily as humans can. For example, autonomous vehicles "see" the surrounding setting, recognize stop signs, and make the decision to apply brakes. AI also drives robots in automated assembly lines and warehouses as well as commercial tasks such as dishwashing and fruit-harvesting.

In addition to flashing out the AI waves, Li has also given his view on how the US and China have fared in these waves. The ability to amass and access a huge amount of data has given a slight lead to Chinese technology giants over their US counterparts. By contrast, in the area of Business AI, China lags substantially by a wide margin behind the US due to the fact that Chinese companies have been slow to adopt data warehousing and enterprise applications. Hence, the cleansing and structuring of data for AI processing will be slower in China. As for Recognition AI, China is taking the lead because its citizens are much less concerned with protecting their privacy in public spaces (and are often happy to exchange privacy for convenience), and China thereby is able to collect and use a huge amount of data. We have yet to see who is going to lead the autonomous AI wave. It may end up being co-led by the two powerhouses.

Some of you may be asking where Europe is in this AI contest. Unfortunately, according to Li, Europe doesn't have a good chance to take even a "bronze medal." This is because, despite a strong tradition in advanced manufacturing and industrial R&D, Europe has only a few of the success factors of the US or China. Europe has never built any successful consumer internet companies, social media companies, or huge mobile applications. Europe also lacks the VC-enterprise ecosystem.[100]

These comments may come as a bit harsh and unfair or even unreasonable. But being an AI company that operates out of the UK, we find that even though AI businesses are thriving and achieving excellence in various pockets in Europe, it is impossible to dismiss many of these concerns.

HOARDING AI: FAANG AND BAT

Another major obstacle towards developing widespread AI solutions in Europe and the rest of the world is that much of the technology and talent is being hoarded by the major industry players in the United States and China. These larger American companies used to be referred to as GAFA: Google, Amazon, Facebook, and Apple. Later the acronym was adjusted to FAANG to include Netflix. In China, the major players are referred to as BAT: Baidu, Alibaba, and Tencent. These companies are excelling in the AI space and leaving small and midsize companies behind.

Let's look at the real-life example of Whole Foods. Whole Foods was a large, but not huge, organic food company. It had a supply chain, it knew where its food was coming from, and it made money selling a large volume at locations all around the country. Whole Foods competed against other companies based on the quality of their products, their location, and their price. It's a traditional business model.

Suddenly, Amazon buys Whole Foods.[101] It can now boost its organization with its proprietary technology, like predictive analytics for customer behavior. It reduces waste resulting from overstocking, and machine learning helps the store determine exactly how much stock it needs to purchase each week.

This indeed shows how competitive the use of analytics will be for Whole Foods. An ocean of consumer preferences (and their behavioral traits), coupled with the optimization of their supply chain, will make it hard for any other traditional groceries companies to compete.

This real-world scenario leads us to wonder what would happen if Amazon were to start to think of itself as a forthcoming and prominent player in the banking or insurance industry. It would be able to impact and disrupt the incumbents rapidly, given the enormous advantage that machine learning can build, within prospects of accuracy within predictions. We are confronted with news of giant acquisitions for billions of dollars daily, which shows how companies that have invested significant amounts of their resources into advanced technologies can afford to play strong in any industry, anytime they wish.

The internet has changed the landscape of how companies compete against each other. Prior to the internet age, we had major technology companies that developed their (usually physical) products in-house, and they were very hard to compete with. Then the internet changed everything, and we saw the rise of powerful tech giants that could compete from anywhere in the world. We see companies highly rich in cash given the efficiency to their cost structure and marginal costs and their cash. In our previous book, we were able to show how, once technology development shifts assets from physical to digital, exponential growth is to be expected, when compared with to traditional companies.[102]

These tech-savvy companies invest heavily in new technology. Smaller AI companies with new ideas and products are quickly acquired and incorporated by FAANG and BAT. A major reason that we haven't seen the rapid commercialization of AI solutions, or the democratization of AI, is because these companies voraciously snap up every new development. As a result, these already large and technologically advanced businesses are expanding rapidly and leaving smaller to midsize companies ever more behind. We will see how these very same companies are now determining both the demand side and the supply side of talent and how this talent is becoming scarcer now more than ever.

These larger companies are preventing democratic access to AI. When they buy AI companies, it's not to commercialize and distribute their products; it's to direct the technology for their own uses. Often, they use new tools to improve themselves internally and ensure that their productivity escalates rapidly.

Additionally, in the same way retailers become smart with smart technology, production can become smart, too. Processes that were simply mechanically automated become predictive. Now production becomes smart production that is self-regulated, learns from its mistakes, and can recognize patterns and make predictions. Just like Amazon does with retailing, it can be done with production to create smart factories. It is automation that is the precursor or prerequisite for machine learning to happen.

When you look at things like smart hospitals, smart houses, smart cities, etc., it all implies digitalization and automation. That's creating the potential for these spaces to become an AI platform through their digital infrastructure. AI can then be made a prerogative. Those companies that have transformed digitally are able to take advantage of this technology much faster than other companies, who will be stuck playing catch-up.

RESEARCHER TALENT IS SCARCE

Usually, when we work with clients to build and integrate AI into their businesses, we have teams of four people: a researcher, an engineer, a project manager, and a quality control person. Hiring a team of four people can be expensive, in large part because people with the skills necessary to do this kind of work are rare.

Currently, there is a shortage of researchers—people who design the AI models based on the clients' needs. There aren't enough people with the necessary skills to build customized AI products for businesses, and as a result, there is a huge gap between AI developments and the adoption of AI in businesses.

Day in, day out, you may hear about amazing AI developments that can achieve a lot of great things. Yet the cost of taking up AI remains high. Even at the trial stage, when it comes to embedding that technology into organizations, it's a lot of work, and it takes a lot of money. Not least because there are not enough research talents available in the market.

This shortage occurs for several reasons. First, there aren't enough AI-proficient graduates out there yet. Training AI researchers takes time. Universities are not churning out computer scientists fast enough to meet the job market demand.

Second, AI professionals are often poached and offered more money by larger firms. Researchers and employees at smaller companies are constantly being snapped up by larger entities like FAANG and BAT. There is a war for signing talent right now, sometimes waged through mergers and acquisitions. If you want a team of AI staff for your company, you may as well buy an AI startup to obtain all the human assets. Smaller developers are made offers that are impossible to reject, and it prevents the organic growth of smaller AI companies.

This leads to some interesting questions about the future of AI technology. In the future, perhaps governments will need to acquire and develop AI technology to level the playing field. Otherwise, they may be forced to depend on private companies to meet their needs with proprietary technology, which leads to all sorts of potential conflicts of interest. Countries that can afford to implement AI will push ahead, while those that find it expensive will lag. In chapter 6, we lay out our arguments for why governments need to invest in AI aggressively and champion its development.

BUSINESSES CLAIMING TO DO AI

For businesses, Big Data used to be the key technology to be had. The explosion in the use of labeled data has led to the creation of many companies that thrive on Big-Data-related services.

Lately, Big Data seems to have given way to AI. After hitting the highest in 2015, the mentions by company executives of the term "Big Data" on earnings calls has begun to decline, according to an analysis of ten years of earnings calls transcripts from more than 6,000 public US companies. AI has entered corporate consciousness, and the use of the term has skyrocketed since mid-2016. In Q3 2017, "AI" and "artificial intelligence" were mentioned 791 times on earnings calls, a 25 percent increase year-over-year. This dwarfs the 300 times of "Big Data" at its peak in 2015.[103]

Understandably, to ensure their economic survival, many Big Data companies are rebranding themselves as AI outfits to get into the attractive market space. And it is commercially a sensible thing to do, considering the growing appetite around all things AI. More and more companies are claiming to "do AI," whether they truly do or not. They may not have the correct qualifications, but that doesn't seem to stop them.

We are seeing a good deal of companies in the market pitching themselves as experts when they are not. We are not alone subscribing to this view. It has been observed that there is a growing number of unscrupulous business outfits that are happy to sell AI-related myths in exchange for profit.[104] A recently published report points out that as many as 40 percent of AI companies don't really have or do AI—they just want to take advantage of the AI hype. [105] Sometimes from the clients' perspectives, it's difficult to know who to turn to for the correct AI expertise. It won't be long before the debate between real AI versus fake AI will emerge or at least move from the inner circle of the tech ecosphere into the big world of public opinion and media networks.

BESPOKE AI

One might imagine AI consulting, where people swing in and build AI for a company, would be big by now, but it's not. The fact is that AI capability must be built from scratch and highly customized. Even the simplest forms of AI require a lot of development and therefore costs.

There is currently no such thing as plug-and-play AI. At least not in the business AI space. Sure, there are plenty of AI models that are freely downloadable out there. Help yourself. But our experience is that they are only around 30 percent accurate—very far off from what is usable in the business settings. This is particularly the case when dealing with the data that is not structured. This setup usually requires humans to check the output and rework the data input.

In the context of business, AI is essentially an integration process that happens inside a specific hardware where people store their data or run their programs. When we work with companies, we develop a solution, the solution becomes a code, and the code is then executed through a program. The program becomes part of a whole series of processes where we are integrating directly inside of a client's infrastructure. Consequently, as it stands, AI as a business capability must be built bespoke.

Like other business technologies, AI projects usually start with a proof of concept or POC. This effectively involves building a minimally viable product to test the feasibility and viability. It also maximizes the learning of the clients' requirements. In order to build the POC in a safe environment, the data used usually is not live, and the AI model is detached from the client's IT system.

Once the POC has reached a satisfying level of performance, the AI model is then integrated. This is the more labor-intensive and expensive part. This process is easier if you are a large company with your own team. Unfortunately, most companies cannot really afford to have dedicated IT and technology teams to deal with the integration and implementation. This means slower progress and takes longer—and more money—to put AI to use.

DEMOCRATIZING AI THROUGH PODDER.AI

Unless the above discussed issues are addressed quickly,

AI will likely, in the near term, continue to be prohibitively expensive for most countries and businesses. Every technology must reach a certain tipping point before widespread adoption, and AI is currently teetering on that edge. If there were a way to make AI adoptable and affordable on a large scale, by offering AI solutions in a broader format that could be customized for unique business needs, that would push it in the right direction. If the AI technology were cheaper, more accessible, and less complicated to implement, it'd be much easier for businesses to start adopting it.

These factors are what drives us at Nexus FrontierTech to build AI that can be standardized and used more readily by small- and medium-sized enterprises as well as independent AI developers. Our aim of unlocking the power of AI for many would have to take the shape of a plug-and-play solution. This is why we have developed and made available podder.ai.

Think back to the early days of the internet. Building a website was a major undertaking that only experts could handle. You had to code each feature line by line. It's difficult to believe now, but it was an arduous and painful task that tested the patience of many developers. Fast forward to the present and it's all drag and drop thanks to programs like WordPress and SquareSpace. These easy-to-handle interfaces and platforms changed the entire landscape. Websites for a business used to be an expensive service, but now, everyone and their grandmother can build their own websites.

The evolution must and will happen to AI as well. The aim of podder.ai is to offer the services, tools, and workflows to both AI developers and companies taking on AI, making it easy for them to work together. As shown in Figure 4, podder.ai effectively serves as a backbone of a company's AI applications system. It is a container-based cloud AI platform, with the flexibility to switch to a hybrid of cloud and on-premise. Since the core architecture sits in the cloud, the platform is not a physical product but rather a service, leading to what we called AI platform as a service, or AIaaS in short.

Podder.ai - where we fit

Designed to allow ML/DL engineers to seamlessly integrate any models without hassle

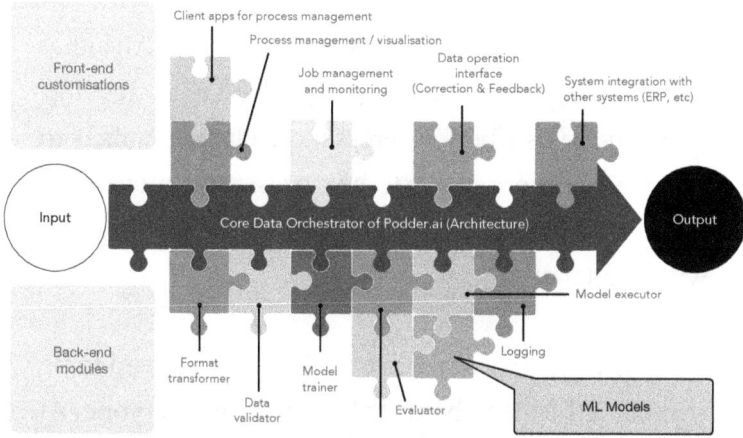

Figure 4. Podder.ai

On the one hand, such an arrangement should be good news to AI developers and researchers as there will be no need for them to expend time, energy, and money on integration issues, thereby freeing them up to focus on the AI models to build for their own clients. Naturally, all the code they write for their models is obfuscated and protected from their clients. On the other hand, companies intending to scale up and upgrade their AI capabilities over time will find podder.ai allows them to work with different third-party developers to come up with AI applications to achieve specific business objectives. They won't have to worry about the problems arisen from the integration of the newly developed applications and IT systems and other existing applications.

To us, there are only two ways to democratize AI: either the government needs to get involved with development and redistribution across more places (unlikely to happen), or there needs to be a grassroots movement towards AI becoming a product, like what we are doing with podder.ai.

Naturally, there are still a lot of issues that need to be overcome for AI to be truly beneficial to everyone. AI will continue to reflect the inherent biases of those who program it, for example. Yet, enabling more people to do AI work with greater ease would inevitably involve more people of different cultures, colors, religions, sexual orientations, which in turn could reduce the data bias.

There is curiosity about AI everywhere, but widespread applications have not yet happened in the past decade. However, with more organizations transforming their productions from nondigital to digital, it will likely only be a matter of time before implementing AI becomes more common. How businesses can do it is what we turn to in the next chapter.

CHAPTER 5

———

APPLICATION OF AI IN BUSINESS

For a while, Big Data was THE hot trend that everyone wanted to be a part of. Traditionally, if you are a company that has a lot of data, you'll have an advantage over competitors for the simple reason that you have more resources to capitalize both Internet and Business AI. Many people considered data the "new oil"[106] or "new coal"[107] in the sense that they are the power sources of most if not all activities. They looked at companies like Facebook who have mountains of data, and they assumed it would make a business better. There is some truth to that because with more data, it's easier to create more effect.

Figure 1 shows a 30,000-foot view of how Big Data can lead our decision-making. One important note is that data collected before the advent of the recognition AI is mostly made up of text. It's obvious that the sheer quantity of data possessed is often seen as the key to business success.

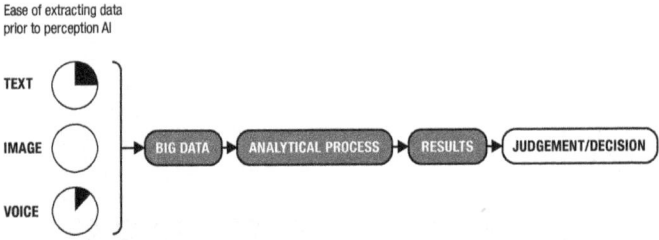

Ease of extracting data
prior to perception AI

TEXT

IMAGE

VOICE

BIG DATA → ANALYTICAL PROCESS → RESULTS → JUDGEMENT/DECISION

Figure 1. Big data - 30,000 Feet View

As mentioned, AI has replaced Big Data as the buzzword du jour. Still, many see data as a critical component of AI. This is understandable, as a large cache of data is often considered a prerequisite for AI implementation—machines need to consume quite a lot of data during the learning process. But data is only one of the ingredients in putting AI into a business. There are other human and organizational factors that must also be taken into account. It is easy to think technological prowess is the key to success in building AI capability and that the rest merely plays a supporting role. Yet, in our experience, the reverse rings more true. There is no question that the basic technology has to function at the expected level, but the corresponding workflows and processes must be altered properly in order to enable the technology to provide the most benefits. Everything needs to work together as a coherent whole to attain a capable AI system.

It's easy to talk about AI in terms of what it's capable of on an abstract scale, but when it comes to integrating the technology into your business strategy, things can get hazy. Corporate adapters are slow and cautious about AI, partly because of the lack of talent discussed in the last chapter but also because of a lack of proper understanding and the costs.

One of the major issues that comes with applying AI to business is that most people have fundamental misconceptions of what the technology is and isn't capable of. There are some jobs that can be replaced by AI, and some that simply cannot. When defining these tasks, you should be extremely purposeful and clear about what you want it to do.

We've created five action points you should consider when determining your AI needs:

1. Be narrow-minded
2. Weigh the risk
3. Get the "last mile" right
4. Consider less data may mean more
5. Do the necessary homework

These are all important points to improve your company's chance of success at adopting and implementing AI.

1: BE NARROW-MINDED

A lot of companies embrace the idea of AI without making the effort to find a clear objective for employing new technologies. Many people imagine AI as a magic fix to social and business issues. As mentioned earlier, politicians are often the most inclined to have complete faith in precise technologies designed to tackle and diffuse social problems. This is obviously the wrong way to approach any major change.

If you only have an abstract or holistic view of what you expect AI can do for your business, prepare to be disappointed. People who try to sell you AI strategy are bordering on pushing you a scam, simply because no one piece of AI technology is going to reinvent your company's strategic goals. We have seen articles written by consultants stating that companies must think about deploying AI to build business strategies—presumably through their services. Such a claim makes very little sense for two reasons. First, just because you've discovered some shiny new technology, you shouldn't completely overhaul the direction, purpose, or major functions of your company.

More importantly, AI at this juncture is most effective in dealing with very narrow and well-defined circumstances. It's important to narrow your scope when thinking about what you would like to use AI to achieve. It is also paramount for you to know the exact business objective you want to achieve. Cutting document processing time by half is a clear goal that AI can potentially reach; using AI to recruit the right people or to sell to more customers is a bit broad and in need of more focusing and sharpening.

One way to think about this is shown in Figure 2. The strategy of a company is only coherent and effective when various properly designed activities are working together (e.g., different parts of a value chain). These activities, on the other hand, are made up of different jobs that people undertake. The jobs that most of us do involve a range of tasks. We do a good job by completing them correctly in a timely manner.

Given that current machine and deep learning systems can only do one task well (e.g., if it is designed to read human faces, it cannot identify anything else), it is much better to think in terms of the tasks, not the jobs, that AI can take over. This is also the reason why machines will in most cases only *partially* eliminate our jobs unless the job is made up of only one machine-replaceable task.

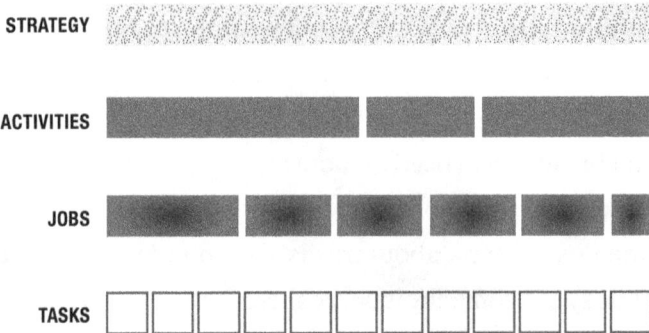

Figure 2. Strategy, Activities, Jobs and Tasks

For example, if your business is a bank, it will be a very long stretch to replace your customer service representatives with AI machines. Serving customers is a job, not a task, and we're simply not there yet. However, you could use AI to do many other tasks, like performing analysis of customer profiles to recommend the right financial products or scanning and processing documents.

A lot of banks require two pieces of official identification to onboard new customers. AI can be trained to locate information like addresses and issue and birth dates on the submitted documents even if they are in different formats (e.g., account statements issued by banks vary a great deal). AI is good at improving in areas where humans tend to make (potentially expensive) mistakes. The so-called fat finger error—wrong keyboard input or mouse misclick—caused Samsung in April 2018 to accidentally issue shares worth some $105 billion to more than 2,000 employees who are part of its stock-owner program. They were entitled to receive a dividend total-ing 2 billion Korean *won*, but were mistakenly issued 2 billion *shares* instead.[108] Around the same time, Deutsche Bank inadvertently transferred €28 billion to one of its outside accounts due to the same error.[109]

Decreasing error *and* speeding up the otherwise labor-intensive is where AI will bring the most economic value. Deploying AI in this case would enable companies to cut costs and raise efficiency. And this is a much better way to invest in AI capabilities. Contrast that to spending a lot of money to build an AI customer representative, which may or may not end up saving you money in the long run, depending on whether your clients can accept interfacing with a machine.

This is where we always recommend our clients start with AI: as a tool to reduce costly, labor intensive tasks. It is far easier to establish—and attain—the expected return on investment with cost savings than with using AI for the purpose of expanding revenue. Start brainstorming these types of tasks, especially if the results of the change are easily measurable. You want to define concrete met-rics that will help you achieve your objective. Using AI to improve accuracy and speed in labor-intensive areas of business is not just good, low-hanging fruit; it is also a quick win to create evidence to your next ups for expand-ing AI capabilities in your company.

2: WEIGH THE RISK

When it comes to AI, humans are going to have to get comfortable with relinquishing some control. Once you implement AI into your business, you need to stay in your lane: there are some things you will be responsible for and some things that you need to leave to the machines.

One of the biggest issues people have with AI is the idea of letting machines make decisions for us. It can be a scary idea, but it doesn't have to be all or nothing. If the decisions are minor, and the machine proves its accuracy, you may be comfortable letting it continue autonomously. If the decisions have major repercussions, then you will probably want to have humans involved in the decision-making with AI assisting by processing data in a way that helps inform those decisions. You are always walking the line between the benefits of offloading those decisions and the risks of an error occurring (of course, humans are not immune to making errors, either).

Some car insurance companies are using AI to assess the damage caused by car accidents. Customers take pictures of their vehicles, send them in, and the AI can come up with the damage-specific repair cost estimates much faster than traditional insurer adjusters who currently assess auto damage. As a result, we will likely see fewer and fewer claims adjusters (together with other insurance employees) in the future, but they won't disappear completely. AI decision-making here works well when the amounts being paid out are small; bigger claims usually require some human intervention. The insurance companies will have to be comfortable with the trade-off between machines potentially making wrong decisions versus the efficiency and cost savings to be had.

AI is now gradually being taken up by law firms, namely to work on standardized contracts and agreements. The opportunity here is huge. Instead of a lawyer taking two hours to redraft standard contracts with various customizations, machines can streamline the process to produce five such contracts in the same amount of time. The risk here is that if a machine makes a mistake, it could lead to litigation. So, even if it is 99.6 percent accurate, that 0.4 percent margin of error may still not be good enough because the associated losses could be huge. Therefore, generating contracts should not be the sole purview of AI; humans still need to check the final product before disseminating it to clients, on top of understanding and being comfortable with the associated risk.

Deciding when humans are needed and relinquishing control over what machines can take care of is a huge part of deploying AI into organizations.

3: GET THE "LAST MILE" RIGHT

This concept builds on top of the prior section. A term widely used in the telecommunication sectors, the "last mile" refers to the final leg of the telecommunications networks that deliver services to end customers. These days, supply chain and the transport sector use the term to describe the movement of goods from a high-capacity freight station or port to their final destinations, such as stores, restaurants, other businesses, etc.

Many of us know how dealing with the last mile is a boon. Imagine you get off a long flight and take the airport train to the city center. To get to your final destination, the choice used to be between expensive taxis or at least one trip on trains or buses, if not both. This is why, to many, ride-hailing services like Uber and Lyft can make the last mile much easier, so much so that it can improve the overall experience of the entire journey.

In a similar fashion, the last mile, to us, refers to the idea that, even with 99 percent of a job being automated, there will always be 1 percent that needs to be handled by humans. There are three reasons why it is important to think through the last mile carefully and how it should be properly integrated into workflows and procedures.

First, we need to create a human failsafe. AI has sparked the debate in what technology can achieve and what will be the potential role of humans in the jobs that we will develop in the future. Increasingly, in some cases, technology may be able to push those to 100 percent accuracy, reaching optimal performance. But right now, however, even the best AI technology for simple tasks would not yield perfect accuracy. We still need a workflow designed around inevitable human intervention.

For the standardized legal contracts mentioned above, law firms would still require human staff to verify the quality of the output by the machine. For even the most straightforward insurance claims, it may still be worth having a person to spot check the payout details. A lot of jobs in airports have been modified or changed as technology becomes available. Take passport control in Europe and the US. In this case, a customs officer may not always be needed to check every passport. Instead, they may be scanned by AI, and if the machine doesn't accept the documents, then a human takes over to inspect them. Now imagine the opposite: you use an e-gate to enter a country, but the machine rejects your travel document and there is no customs officer available to examine it. This would leave you in an impossible situation. We also see this shared responsibility with the self-checkout kiosks at supermarkets and pharmacies. There is always a need for a human employee to oversee the machines and assist customers.

Those are all examples of highly repetitive jobs shifting

in nature to reduce the workload of humans. These aren't the dark factories that function entirely on their own from smart automation down to the sensor light AI; these are customer-facing roles. Which is why we still need some sort of human involvement.

The second reason is that, while certain things can be automated, there are still a lot of jobs that are in our best interest to leave to humans. For example, hotels are a service industry that is mainly dependent on the customer experience. A bunch of impersonal robots checking people in and caring for their needs is not going to produce the same warm effect that human staff members can. In a recent conversation, Mark learned that a large hospitality group in the UAE introduced a hotel check-in using only robots, and guests ignored them entirely. People preferred waiting in line to check in with a human over using the robot. Just try asking a robot for a room upgrade. For a variety of cultural reasons, the same technology that works in airports doesn't translate to hotels.

Thirdly, there are tasks that are simply impossible for machines to take over from human beings. Have you ever wondered why, after you have obtained your boarding pass from a machine, you still must interact with human staff to check in your luggage? This question had puzzled Terence for a long time. So one day he decided to find out why by asking the attendant behind the counter.

The answer, as it turned out, was very simple: their job is to attach the right bag to the right passenger. That's it. Presumably, otherwise, anyone can check in a bag on any flight.

Some of our readers may have a different experience. When Mark and Terence were going through the new terminal at a Singapore airport, they didn't have to interact with any human staff when checking in their bags. Instead, they left their bags in a tunnel-like machine and scanned their boarding passes on it. After that, their bags were transported away by the belt inside. It may seem like humans were no longer necessary, but this isn't true. What this new process did was transfer the task previously conducted by airport staff to passengers, in this case to poor Mark and Terence. Humans are still required.

There's always one little niche that robots cannot fill, and we need to have humans working with them to get things done. The last mile is a great argument for policy intervention to make sure we maintain this attitude towards the importance of people. The last mile idea also points to the fact that while machines are taking certain jobs away from humans, they are also creating other opportunities that don't currently exist. We need to create a more cobotic-friendly environment where humans work alongside robots in a directorial role.

4: CONSIDER THAT LESS DATA MAY MEAN MORE

As mentioned before, data and AI have a historical connection where a lot of emphasis was put on the value of acquiring, even hoarding, data. Most companies believe they need lots of data starting out, which in many ways is true if you're looking to train a model or machine. Big Data also allows people to glean a lot of insights from gigantic sets of data. A lot of people believe that if they have enough data, they can create a "data moat"—a competitive advantage that is conferred by having proprietary data—using large data sets to do more than their rivals.

Nonetheless, the idea of "the more data the better" is a misconception. Ultimately, when it comes down to how much data is needed to gain a competitive advantage, it really depends on what exactly you are trying to do.

HOW MUCH DATA DO YOU NEED?

There are three thresholds to consider when deciding what and how much data can really do for you. This is the reason why the oil analogy is flawed—data is not the new oil.[110]

The first pertains to minimum algorithmic performance, or MAP. It's like the concept of a minimum viable product (MVP), where companies try to create the most basic version of a new product to test before moving forward.

There is no need to spend time on the bells and whistles. Instead, the focus is on demonstrating the performance of its critical duties. In these situations, people push out MVPs to let the potential customers decide if it is good or not and collect feedback where improvements should be made. For AI, the same analogy can be drawn; there is a need to understand the minimum level of accuracy that will be needed to achieve the designated goal. This is described as the MAP threshold.

When it comes to things like getting machines to read legal contracts and X-rays, you probably want the MAP to be as high as possible. Results of low accuracy deem the algorithm worthless (e.g., failure to detect cancerous elements on X-ray photos) if not downright dangerous (e.g., litigations resulting from mistakes in contracts). In these instances, having more data could improve accuracy.

On the other hand, there are situations where low MAP is harmless and good enough to do the job. For instance, the predictive suggestions you are given when messaging on your phones or Gmail. When it does finish sentences the way we wish, it's a boon. When it doesn't suggest the right words, we don't seem to mind at all; we just continue typing. The machine's wrong guess doesn't matter very much. Having loads of data in this case does not dramatically enrich the results or user experience.

The second consideration is the performance threshold. Whereas there are times when 100 percent accuracy is not needed to solve every problem, other times, problems can be too complex for a machine to solve even with perfect accuracy. In this case, no matter how much data there is, it will not help reach the objectives. And then there are times where the task is easily definable and straightforward enough that it is possible to achieve near 100 percent accuracy with even a small training set. Take driving licenses in the UK. Since they have standardized format across the country (unlike the US), we can easily train a machine to pull data from them without resorting to a large data requirement. The same category of data is always in the same spot on the license and the two dates are easily distinguishable—the older date must be the birth date and the more recent one is the date of issue. In this case, it's quite simple to achieve near 100 percent accuracy. If we can achieve these results using relatively small data sets, then we achieve the performance threshold we need.

The third consideration is the stability threshold. Machine models decay over time because data sets evolve and become outdated. Let's say you sell kitchen appliances and pots and pans online. You have lots of data on these items, but the moment you pivot to sell something else, like linens and towels, the data sets you have will no longer be useful. No matter how big the previous data set was, it doesn't work anymore. The amount of data isn't as important as its relevance to the problem you need to solve. This threshold is an important consideration for driverless vehicles. Can the data from previously trained cars make the vehicle smart enough to deal with all future eventualities? Again, the vast amount of data collected doesn't lead to a better and, more importantly, safer performance. Quantity is not the same as quality.

QUALITY OVER QUANTITY

Having a lot of data doesn't always create competitive advantage. Owning only a small set of data doesn't mean competitively disadvantaged. We have a client in Japan who sells furniture. One of the biggest challenges with furniture stores is people don't buy furniture every day— or every year, for that matter. Since these are occasional purchases, it prevented our client from capturing a lot of data from repeat customers. It asked us to help it find a way to give relevant recommendations to online shoppers, like Amazon does. It wanted to incorporate little reminders that if you're interested in this one chair, you might also be interested in the matching couch, or this other chair.

The challenge was teaching the AI to make these recommendations without a lot of data. Amazon obviously has millions and millions of customers and a decade of buying information to pull from when making those recommendations. Since our client's customer base is much smaller compared to Amazon, we decided to survey customers about their preferences. We could use this smaller data set to train the AI to make effective recommendations. While a company like the furniture retailer would never be able to reach the level of effectiveness enjoyed by Amazon, we believe that, despite a significantly small set of data, we were able to get the company's recommendation system to be 70 percent as good as that of the retail giant.

The fact is that today's AI technologies have advanced enough to be able to achieve goals that preclude the need for plenty of data. In many ways, what matters more, indeed much more, is not possessing "Big" Data but rather the "right" data. AI models can only attain their designated and desired objectives if they are fed with the right sort of data. As an illustration, imagine we are building an image-recognizing algorithm that can identify Asian faces based purely on the skin color in a multiethnic community, in which the members among them have, say, a total of ten different of skin tones. (In practice, AI often relies more on the unique features such as distance between the eyes, and nose shapes, in addition to skin tones, to recognize faces.)[111] The key to training the algorithm to increase the accuracy is not about having a lot of data points: 10,000 images for only eight out of the ten shades of skin tone will never allow the model to be fully attuned to the ten different tones across the entire color spectrum. It can't recognize the two skin tones it hasn't been exposed to at all.

In this case, instead of blindly pushing some 80,000 pictures into an algorithm, it will be far more effective and useful to train the machine by feeding it with images covering all ten clusters, even if there are only one hundred available for each cluster. Put differently, having a small but representative set of data is definitely better. AI will never work as intended with a sizable but biased or incomplete dataset. Small but the right kind of data is always champion. What the above discussion has revealed is that it is possible to achieve results with limited amounts of data. In fact, we are getting better at using limited data sets to make accurate predictions every day. This idea pokes a hole in the idea of Big Data as the new oil. And this argument may only work in the eras of Internet AI and Business AI, when owning a large amount of text data is beneficial. These days, however, with Perception AI, a far greater variety of data, such as image and voice, can be rather effectively compiled. Figure 3 shows an AI version of Figure 1.

Ease of extracting data
brought about by perception AI

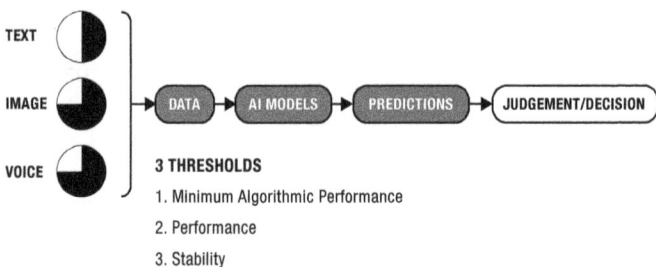

Figure 3. AI - 30,000 Feet View

An observation that is interesting is that existing technologies allow machines to recognize images and voices to the extent that surpasses the ability to deal with text. This doesn't mean machines are incapable of working with natural language and text. The evolution in this area has in fact been moving very rapidly. In some ways, AI is effectively democratizing the extraction of data.

ALTERNATIVES TO BIG DATA

Indeed, new techniques are being developed to compensate for the inability to collect data. We will briefly mention two here.

One late development in deep learning is generative adversarial network, or GAN. These systems are deep neural net architectures that are comprised of two nets, pitting one against the other (hence the "adversarial"). One of them is called the *generator,* which generates new data instances, whereas the other, the *discriminator*, evaluates them for authenticity. The discriminator must decide whether each instance of data it looks at belongs to the actual training dataset or the fake ones produced by the generator. Without getting mired in detail, one of the greatest benefits GAN can provide is constructing synthetic data. In other words, we can use deep learning to create new data to train other deep learning networks, effectively making up for the lack of data.

Another development in AI that aids with limited data sets is something called *transfer learning*. In this approach, we are getting a neural network model designed for one task and repurposing it for a second. Devices like Alexa and Siri require so-called wake words to trigger their activation. The problem is that machines will have to be able to detect the wake words spoken in different accents and tones. It would be a lot of work to find enough different people and record them saying "Alexa" or "Siri."

One way to overcome this seemingly insurmountable challenge is not to train the machines to understand the wake words. Instead, what is now possible is to train the system to learn spoken language using data you already have. Only in the last layer of the deep neural network is the wake word introduced. In this case, the lack of data is not an inhibiting factor for developing AI models.

5: DO YOUR HOMEWORK

We can't stress enough that AI can only do very specific, well-defined tasks. Because of this narrowness, if you want to put AI to effective use, you need to make sure you know exactly what goal you're trying to achieve as suggested in the rule "Be Narrow-Minded." You must do some homework.

A lot of the time, customers we deal with don't know what they want or only have an abstract idea of a goal that they have in mind. They know that they want to reduce costs, but they don't know how to go about doing so. AI isn't built to serve abstract purposes. You need to know exactly what you want to get a model to help you. And this is a question that only you can answer.

But this is only a starting point. Next, you should sit down and map out your current workflows and processes. This is because, as shown in Figure 4, technology must be supported by the right workflows and processes to maximize its potential. In turn, the workflows and processes should be backed by staff both managerial and IT.

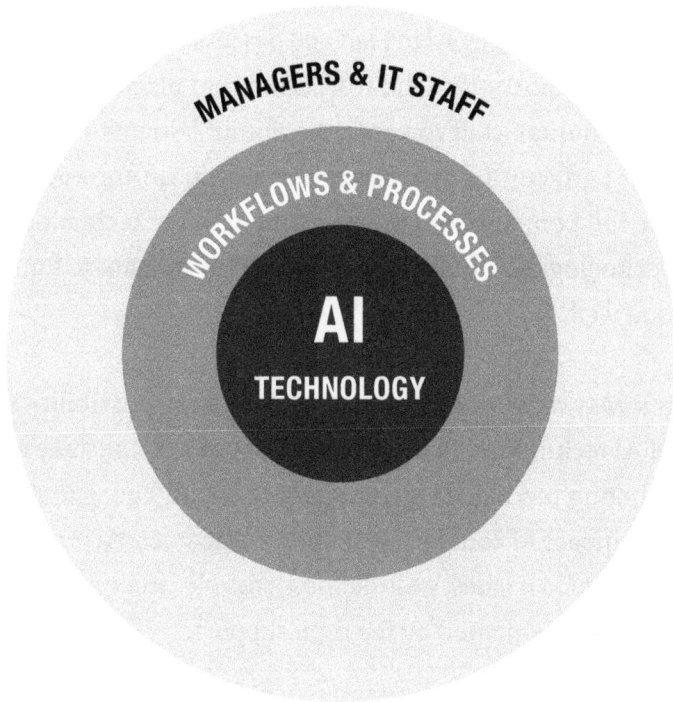

Figure 4. Supporting AI Technology

This may sound like we are stating the truly obvious. But we have been surprised to see that clients' enthusiasm in taking on AI wanes as soon as they are asked to lay out the existing workflows and processes they seek to improve. As a matter of fact, after determining what tasks can be solved by AI, it is necessary to come up with a new set of processes and workflows to accommodate and dovetail the newly-developed technology in place.

Recall our earlier discussion on how human and machines tend to work best together. Imagine a situation where you have started using AI to read and process documents. Yet you have not gotten the workflows that place the "last mile" human staff to verify the output. No matter how near-perfectly the machine performs, you are not reaping the full benefits from the investment in the technology. Technologies must be fully supported by humans to function well.

It's easy to be overwhelmed by the sheer possibilities of AI technology. Many companies make the mistake of thinking too big. They overgeneralize and exaggerate the impact AI could have on their companies. AI is most effective in narrow, well-defined, specific circumstances, and complemented by the right support.

When approaching AI for your company, it's important to stay grounded in the goals and missions of your business. AI cannot define or replace your business strategy. AI for AI's sake is not useful or cost-efficient. But employing AI strategically to help advance your business can be a boon for everyone.

PART III

THE IMPACT OF AI

CHAPTER 6

HOW GOVERNMENTS SHOULD PREPARE FOR AI

Many businesses and institutions are unprepared for the rise of AI, but governments may be the least prepared of all. Although AI has existed in some form for several decades, the major governments of the world have mostly been silent and uninvolved. In that void, AI has become the purview of giant for-profit companies to the extent that the industry is driven almost entirely by the private sector.

This widening gap raises a slew of ethical questions about the future of AI. We believe the government has a responsibility to become a part of this dialogue, to help even the playing field between private and public sectors and prepare their citizens for the consequences of new AI developments. Countries where governments refuse to get involved in furthering this process along risk losing competitiveness in the world market, especially as AI becomes more and more accurate and advanced.

This chapter will highlight the countries that are getting it right, the drawbacks of implementing AI in the job market, how AI may impact society inequality, and the potential creation of a "useless" class. While AI is potentially a remarkable tool for change and innovation, there are many potential pitfalls, which governments should play a role in helping society navigate.

GOVERNMENT RETICENCE

There are several reasons that many governments haven't yet stepped up to the challenges posed by artificial intelligence.

First is that, by nature, government bureaucracies are poorly designed to deal with the challenges of new technology. They don't respond to innovation very quickly and struggle with adapting to change and legislating new business models. This is complicated by the fact that AI isn't a business model but a strategic technology tool that can be vertically integrated in many diverse ways across different areas. Any government would have a tough time embracing such radical transformation.

When politicians do initiate conversations on AI, their focus is usually on jobs and AI's potential effect on employment numbers. This is where the bogeyman of the job-stealing robot rears its head again. No elected official wants to support technology that steals the jobs of their constituents. It's bad for keeping their own job, as well.

We've discussed before why fear of potential massive job losses is misguided, but coming from elected officials, it's also extremely shortsighted. One of government's main challenges is to revive the economy after a recession. Most of the time after a downturn, we see the creation of new productivity and efficiency gains that contribute to GDP, lifting the economy above where it was before the crisis to a new level of prosperity. In the last couple of decades up through today, however, it now takes much longer to recover from crises, to the point where some sectors never recovered after the Great Recession of 2008, despite government stimulus. So if extensive use of macroeconomic policies and subsidies aren't doing the trick, what will?

We think that the introduction of AI is an opportunity to boost the economy by generating new markets that don't exist yet. This is because AI illuminates areas of possible economic intervention and recovery that we have never seen before and have yet to predict. Along with this comes the possibility of new job creation. For example, consider the possibility of using AI to improve on cancer prevention procedures. That opportunity could lead to more net employment due to research grants, internships, funding for PhD students, and operational labs.

We have repeatedly seen that job creation happens when the integration of new technology and discovery leads to commercialization. Entire cottage industries can also be built on a single technology. For instance, the development of smartphones created a new hardware category that needed entirely new operating systems. As a result, Google Play and Apple's App Store were created, and mobile app development became a whole new class of software. To date, there are about 2 million apps available for Android OS and iOS.[112] Not only that, but mobile app revenue reached $92.1 billion in 2018.[113] Smartphones are ranked up there with vital life support and essentials. Indeed, for many refugees, smartphones are considered more important than food and water.[114]

Another prime example of this is the space industry. At first, these initiatives were government driven under the umbrella of NASA. Now, companies like Blue Origin (started in 2000 by Amazon founder Jeff Bezos) and Elon

Musk's SpaceX are jumping into the playing field and creating new jobs. As a well-funded company, SpaceX had over 6,000 employees in 2017 and has had more than 1,000 job openings since then.[115] The goals of space travel technology have expanded beyond scientific study and exploration to the commercialization of space travel for tourists, and getting to Mars. As the young industry grows, many more new jobs may be created in the future. Governments need to be reminded again and again that history has shown that new technology helps create jobs, not destroy them. If they can be taught to trust this reality, perhaps they will be ready to encourage and support new technology like AI.

EVOLVING JOBS

As discussed throughout this book, AI is best utilized by replacing humans in repetitive jobs. Instead of merely eliminating these old jobs, the result is new opportunities with increased flexibility. People who were once doing tedious labor are now supervising, checking work, and making decisions. These positions can adapt because they have an intrinsic level of agility in how they might evolve.

Consider the rise of freelancing as a career. Originally, freelancing was a part of the service economy, where individuals met the needs of businesses by providing expert services the company did not possess, like graphic design or proofreading. Then the trend moved towards more project-driven opportunities. Companies would bring in freelancers to serve as the experts for the projects they were running. We then saw an increase in the use of expert consultants, an entirely new form of freelancing than what we started with.

The same pattern that inspired the rise of consultant hiring can carry across industries. With the rise of AI, we will see a decreased dependency on the traditional industry designs of the nineteenth and twentieth centuries, with their emphasis on industrial methods or mechanisms. Those methods were systematic by nature because repetition was the only way an organization could control the amount of output it was generating. Everything—from the number of hours we work and the rise of unions all the way to salaries, minimum wage, incentives, promotions, and required output—was designed in a model where you had some degree of that systematic repetition despite the process. It's no secret that the most effective factories are what we term dark factories. This is a very different paradigm than before.

In the past, management was nothing more than layers built on top of the different functions that would oversee the traditional value chain. Now that technology is becoming more and more affordable in terms of what is viable, it will impact the managerial needs of those kinds of industries. This time, it won't be people managing people. It will be people managing AI.

Already today, some of the skills in demand are machine learning, deep learning, data science, and predictive analytics. All of these jobs are related to working with AI, and by some counts, there will be a net gain of 500,000 jobs by 2020 (1.8 million jobs would be eliminated by AI, but 2.3 million jobs created because of it).[116] And right now, there aren't enough people with the right skills to do them.

When it comes to jobs in the new AI paradigm, governments need to sit down, observe, and respond. They need to be a guest at the table on this topic of conversation, not only so that they can understand the potential for new job growth (instead of just loss), shape it in an ethical way, and create training programs so that their respective countries stay competitive but keep their own viability in a technological future. Today, governments are already having problems competing for cybersecurity experts and computer programmers. They have no hope of even keeping pace with society if they don't try harder to change the pattern soon.

FRONTRUNNERS IN AI

Although the general trend is a lack of attention toward AI in politicians and legislators, there are a small number of national governments that stand out for the proactive stance towards exploring the possibilities posed by AI.

Countries like Canada, France, the United Kingdom, Germany, Denmark, among others, have all started initiatives to explore the ethics and responsibilities surrounding AI. The International Panel on Artificial Intelligence (IPAI) was announced by Canada and France in 2018 as a platform to discuss "responsible adoption of AI that is human-centric and grounded in human rights, inclusion, diversity, innovation and economic growth."[117] There is a hope that other developed countries will join IPAI in the future.

The previous year, the UK government established the Centre for Data Ethics and Innovation as an advisory board, with the intention of exploring how to support AI and Big Data projects in an ethical manner. Germany took a reactive stance, focusing specifically on self-driving cars. Seeing that widespread adoption of the technology is inevitable in the long run, it created an Ethics Commission on Automated Driving. The commission eventually developed a list of twenty guidelines for the motor industry to follow when developing automated driving systems. Denmark and the EU both have earmarked funds for an AI strategy to build the economy while maintaining ethics guidelines that address concerns like fairness and transparency.

We see a growing number of governments beginning to tentatively explore these possibilities, but none of them is as fully fleshed out as what's happening in the United Arab Emirates (UAE). The UAE is the first country to have formally institutionalized a Ministry of Artificial Intelligence. Being that the UAE is a state that runs more like an enterprise than a country, there are a lot of incentives to embrace AI functionality. What sets the UAE apart is the holistic viewpoint of its AI strategy. The official government statement outlines its plans to invest and use AI to improve government systems as well as to invest in the advancement of AI technology to enhance competitiveness and lead the market.[118]

It's also likely that the government sees a benefit in highlighting its forward-looking stance on AI to foreign investors with an official government unit. What's more, it also has initiatives that address the effects on future employment needs by including AI-related curricula in schools, training programs, and summer camps. Omar al-Olama, the UAE minister for AI, drives the long view approach to its strategy and reminds us all that technology has been displacing workers for decades. "It's not just AI," he recently said at a conference in Dubai. "It's important for us as humans to be agile, and people who don't accept change are unfortunately not part of the future."[119]

More than anyone, the Chinese government has

embraced AI technology for multiple applications. It uses it as a tool for monitoring its very large population, employing it in ways that can have huge social ramifications. Take, for example, the introduction of the social credit score in 2014, an individualized good citizenship score that considers a huge variety of factors from purchasing behavior to whether you pay your bills on time. If you have a bad score, you are penalized with slower internet, less access to jobs, and potential travel restrictions. Much of this information is collected and analyzed using AI.

The Chinese government is investing heavily in AI as a tool for public safety by installing facial recognition technology in the 170,000 million CCTV cameras positioned throughout the country.[120] It can use this to track people of interest and even to catch criminals. In one famous case, it picked up a man guilty of embezzling government funds ten years after the fact. He had lain low for a decade but finally showed his face when attending a concert, which was monitored by CCTV cameras. It's almost like science fiction; you can run, but you can't hide from the Chinese government! In a government-approved test of the new facial recognition system in late 2017, BBC reporter John Sudworth tried to evade the technology but was apprehended in seven minutes.[121] China's technology is only improving, and with plans to install another 400 million CCTV cameras by 2021, soon the government will be able to use its cameras to find any citizen in just minutes.[122] In fact, its goal is to be able to identify people within three seconds with 90 percent accuracy.[123]

Machine- and data-based monitoring are not only restricted to China. Use of biometric data is also prevalent in the West, in places like Germany, for safety and security reasons. Biometric data is also used in the US for security. The government has mandated that its visa-exempt partner countries convert to biometric passports if they wish to continue receiving automatic visa waivers.

It's easy to see how the above examples encapsulate many people's greatest fears about the government's use of AI. People are wary of creating a Big Brother type situation (a la George Orwell's *1984*), where the government can see and potentially control everything its citizens do. It's important to remember that, as things stand in general right now, private companies are powering the clear majority of these technological advances, not the government.

In fact, a savvy government might be able to limit the extent to which these private companies can observe, capture, and sell our data. Look at the issues that the US has had with voting machines, from software vulnerabilities that made it possible to meddle with voting records to selling used voting machines with real voting data inside. The US government has effectively stayed out of the conversation rather than addressing these problems, and that needs to be rectified.

European governments have set a better example when it comes to using legislative power to protect private citizens. Proposed in 2016 and implemented in 2018, the General Data Protection Regulation (GDPR) is a response to the widespread power and influence of technology and the new ethical issues that have arisen along with its growth. GDPR requires companies to provide full transparency on the collection, processing, and storage of user data. It also requires companies to be true to their word on how and for what period they will use your data. Government intervention doesn't only benefit private citizens, either. Companies found to be in violation of GDPR can be fined between 2 and 4 percent of their annual financial turn-over. In the case of large corporations, the amount can be substantial. Google, which has had issues with disclosing how it collects and uses personal data, was fined $56.8M USD after GDPR was passed. The advent of GDPR is an exemplary case of how governments can respond to new technology in a progressive fashion, and it's easy to see that governments can play a similar role with AI, especially in the way they can help steer AI's growth in a positive direction without exploiting private data.

The downside, however, is that there will be less data

available to train and advance AI models. There are a lot of people who are willing to trade their data privacy for comfort and convenience or security. This is the case in China, where people are more laissez-faire with the privacy of their data. The result is that the country could keep pushing AI developments at breakneck speed.

PUBLIC AND PRIVATE PARTNERSHIPS

It's time to start seriously considering the nexus between the private sector and the government. We need to think about closing the gap between the level the government is currently operating on and the level that major corporations like FAANG and BAT are operating on.

One way to do that is to establish more partnerships between public and private companies where you have, for example, public assets being managed by private companies. This is a good baby step that governments can use to take advantage of new technology without taking on the responsibility and cost of developing it themselves.

You can often see this in large European cities where public transport is often powered by private companies with public money. It leads to innovation and more effective management. In Madrid, Spain, the bus system in the center of the city is run by the municipality, but there are private "ring roads" that surround the city for buses owned by private companies. The decision to outsource made it easier for the city to quickly provide more service to outlying communities without having to manage the implementation on its own. In Melbourne, Australia, the municipal government also outsourced a bridges-and-roads toll project, allowing the private company to take on the risks related to construction completion and cost overrun.

Building upon these successes, we could take the public–private partnership one step further and imagine governments employing startups to fulfill municipal duties. Private companies benefit not just from monetary contracts but from the opportunity to get their technology into the public sector. It could be an attractive alternative for companies to being systematically acquired by FAANG and BAT, as is common now. This would lead to a much more equitable relationship between the government and the private sector regarding new technology while also allowing startups to continue innovating independently rather than being absorbed into larger corporations. Of course, issues of security would need to be resolved, but the idea is worth exploring.

THE WIDENING GAP

The "American dream" is alive in places like Denmark right now—truthfully, more so than in the United States.

Scandinavian countries in general are heavily digitalized, educated, multilingual, and invested in renewable energy. They are also some of the most innovative countries in the world based on patent activity and the number of researchers. They have been forward-thinking in terms of merging technology with government services, using existing innovations to create a single personal electronic ID code that works for everything, from paying taxes to making medical appointments. It does not only save time for its citizens but also builds up the IT foundations and technological nimbleness of each of these countries. The governments also incorporate security features like two-factor authentication. The adoption, application, and investment in current technology for government services are factors that help create competitive economies by setting an example for the private sector to do the same. It's not a surprise that Denmark, Norway, and Sweden rank highly in terms of standards of living and infrastructure, and it is these countries who may be the ones taking the lead in AI as a burgeoning industry.

It's imperative for governments like that of the US, as well as developing countries, not to fall behind in embracing technology that can boost their economies and standards of living. There is already a gap forming between those that do and those that do not—and that gap is only going to get wider. We always hear politicians talk about losing ground to other countries. Europe worries about falling behind the US, while pundits ask if China has moved ahead of the US. When will action follow the browbeating?

AI is ripe and new enough for governments and businesses to begin thinking about partnerships. Countries can use these relationships to improve their competitiveness and close the digital divide. If systems across the government start to run more efficiently because of well-employed tech, that efficiency improves productivity. With less cost and more efficacy, new systems, like Estonia's e-government model, allow a country to increase its technological readiness and attract new types of business and investment. The businesses that would be attracted are the ones that will be more heavily invested in the digital services, meaning more tech companies and innovation. At this point, you would be bringing in not only new jobs but also the best minds to help develop the jobs of tomorrow. But this type of cascading growth and transformation only happens with a government's political will to do so.

Let's return to Estonia as an example. Estonia is known as an e-government because 99 percent of its services can be accessed digitally and remotely using the same electronic IDs as Denmark, Sweden, and Norway. So why did Estonia's leaders choose to digitize as much as possible? Their answer is very transparent and clear. Because

Estonia is a small country of about 1.3 million people, its best hope to grow as a nation is not through increasing the birthrate but by immigration. To successfully attract people to the country, the immigration process must be easy, as must access to the government's services. In fact, it only takes eighteen minutes to register a business in Estonia, with no paperwork involved. As of 2018, it takes three minutes to file taxes in Estonia, and eventually the entire process will be fully automated. Another feature of Estonia's e-government is that all databases are interconnected, so that if you update your address within one entity, you won't have to update it again anywhere else. This means that if you change your address at your university, your hospital's records will automatically receive the new info and be updated as well. With a "digital by default" approach to running government, it is no coincidence that the IT sector accounts for 7 percent of the country's GDP (as compared to agriculture, which is less than 1 percent). By all accounts, Estonia's e-government strategy to gain more citizens is working. In 2014, Estonia opened its digital government services to the world. In just four years, 40,000 people from 150 other countries became e-residents and registered more than 6,000 companies.[124]

Although public–private partnerships are not going to be as involved in the transactional parts of the economy the way private sector companies are, they will be creating added value. They will operate at the upper part of the value distribution, charging higher prices, and, therefore, making more money and paying higher wages. In turn, these higher wages will make it back into the economy by supporting the industries working inside of it. More disposable income allows for more jobs to exist in the realm of service and luxury goods.

This is the importance of digitalization and how smaller countries can compete with larger, wealthier ones. Governments need to understand how this impacts their competitive positions on a global scale. Countries that have a blank canvas can theoretically leapfrog from zero to a hundred if they have the right leadership and the right investments. Meanwhile, it will take much longer for a country with a long industrial history to eventually adopt and adapt, because it must overcome its attachment to traditional methods.

With the right mindset and political will, however, this can be done. It might come as a surprise to some, but Germany was almost left out of the digital revolution. For a long time, Germany was solely committed to building up its industrial capabilities. It was so distracted that it almost missed the boat on renewable energy and modernizing its infrastructure. Luckily, it pivoted and accelerated its technological development to continue to participate in the conversation.

Many countries in the EU are having a hard time catching up with AI, whereas the United States is better equipped to push AI initiatives. It has a lot of venture capital that allows it to take more risks with startups.

Big companies like Google, Apple, Facebook, and Netflix continue to dominate the market because they got there first. They have an established model and enormous resources. Netflix's budget to make *The Crown* is ten times that of a regular BBC show. Amazon is cornering the retail industry in the US and EU. Anyone who cannot compete will only fall more and more behind in a world where large corporations dominate the entire world market. The same will apply globally. Unless a country has developed policies to push out AI, it's going to be an uphill battle to continuously generate income, raise productivity, create higher value jobs, and raise the standard of living. One can't just say it is going to live outside this arms race. On the contrary, it will probably become one of the larger discriminators among countries.

The gap is starting to widen between the countries that are invested in encouraging a digital future and those that are beginning to lag. It doesn't matter how long you work or how many people you employ. If you're not integrating a digital system, the cost of adopting it later will be extremely high. This is not a conversation that governments can run away from. It's time to engage before it's too late.

WHAT'S AT STAKE?

As urgent as the adoption of AI by governments seems to us, we don't see the evidence of this urgency shared by the governments. Most seem to still view it as a trend that will come and go. It is pessimistic to believe governments won't step up, but the truth is that if they don't, the outcome could be bleak.

One of the worst-case scenarios is one where the government remains absent from the AI space for many years while AI continues to shape the economic and technological landscape. If the government suddenly begins to legislate without really understanding the situation, it could cause a lot of damage. The policies themselves become disruptive, and the government is simply too slow to course correct. Now the government doesn't own the conversation, and the fear of deterministic AI causing damage becomes a little more realistic.

Technology by itself can be disruptive, but bad government policy can be even worse. We've seen the chaos that occurs when bad governing plunges a country into uncalculated change, as seen in Venezuela, Brazil, and most recently with Brexit in the United Kingdom. In Venezuela, the directive to continue nationalizing power infrastructure, telecommunications, transportation, and more in the face of a severe recession created an economic collapse. Brazil was sent into paralyzing stagflation due to poor fiscal and monetary policy set by former president Dilma Rousseff. Based on these situations, we can see how delayed government action that allows technology to be run unchecked alongside bad policy decisions might cause quite a national disaster.

Henry Kissinger wrote an article for the Atlantic in July 2018 where he identified one of the key risks of government apathy towards AI as the potential that AI could mutate in very ugly ways.[125] We previously discussed the possibility of deterministic AI that could make decisions on its own on behalf of humans. This is not a direction we want to move toward: a future where AI makes decisions on what's best for us, removing our free will.

This sounds very dystopian, but Kissinger's main point is that if governments don't begin to own the conversation, then AI could become a powerful technology that creates more harm than benefit, because of society's inability to handle or control it. So far, it's hard to say whether or not that the companies who own most of the technology out there are relatively benevolent. What Facebook did with Cambridge Analytica is very much reprehensible, and Google has been found repeatedly to be collecting our data without our permission.[126] Will companies like Cisco, IBM, Apple, and others continue to stay ethical and use our data properly? Remember, their number one goal is to make money. What if they also suddenly started to behave unethically? There is a danger to handing them too much power. Unfortunately, with the exception of GPPR, governments have no way to protect themselves, or us, from this.

INEVITABILITY OF CHANGE

No matter how much resistance we offer, economics always wins. It makes economic sense for private companies to use AI and smart technology to improve efficiency and be cost effective. Unions may try to fight back, but history has repeatedly shown us that it's impossible for them to fight forever.

We saw this with the sharing economy model, where taxi drivers lobbied hard against Uber. They were trying to protect their jobs from this new, technological juggernaut of a concept. Their protests sometimes resulted in temporary bans, like in Barcelona and Vancouver, but Uber eventually moved back into the space. In places where Uber was suspended, local companies began replicating the same mechanisms. Taxi drivers will have no choice but to adapt or lose their jobs.

This conversation about AI is not optional. Historically, countries have sometimes opted out of participating in technological advances because they were able to rely on natural resources like oil. This revolution is different. In the end, it won't matter where on the maturity scale countries want to be; it will simply come down to either being online or offline. Being offline will have major repercussions in terms of dismantling them from the global supply chain and losing business. This is why this conversation may ultimately become the most universal, pressing conversation of our time. Governments need to feel this same urgency.

Many jobs that have existed mostly unchanged for decades are starting to come to the end of their life spans. This situation scares both workers and politicians. It's not that new technology is bad or evil; it's just neutrally eliminating certain positions. Over time, those jobs will phase out and new ones will be created. This is how society evolves.

Where will these new jobs appear? And how will we train workers to fill them? The way we educate people will also fundamentally shift because the purpose of education, ostensibly, is to prepare people for employability, which we will discuss in the next chapter.

EVOLVING ECONOMIC MODELS

When people are being displaced by technology as it scales, the question then becomes what should the government do about it in the interim?

One idea is to tax the excessive wealth generated by more-productive robots. Bill Gates has espoused support for this idea. You would basically be taxing the excess productivity that companies are generating from the use of automation. The problem with this is that it creates a new tax bracket. Companies will protest and lobby against this. They may decide to move to jurisdictions where it does not apply. With cloud technology, you don't need to be in a specific country to do business; the cloud is everywhere, and machines can operate anywhere.

Nonetheless we believe that taxing robots is an inevitability. If more jobs are being given to robots over humans, then the government will lose the income tax revenue it would have made taxing those employees. This will put governments in a tough position between needing the tax revenue and not wanting to drive corporations out of the country. It will be a difficult quandary to sort out.

Another strategy that is gaining momentum is the idea of a universal basic income (UBI). This idea is being tested out in Canada and several regions of Finland and was the subject of a referendum in Switzerland (where it was rejected). The idea is to pay every adult in the country a minimum monthly salary, unconnected to their employment status.

UBI is a popular idea because people like the idea of the government taking care of them, even if they lose their jobs. These experimental pilots focused on providing UBI for members of society who were unable to work full-time and already receiving public assistance due to disability or health issues. If UBI is truly an option, they also need to be testing the idea with able-bodied workers. Alternatively, instead of a salary, other suggested ideas include a minimum condition, which could be public housing, utilities, and subsidies for food and schools.

The problem with UBI is you can't run an economy on subsidies. A look at the US government's agricultural subsidies with a critical eye points to deleterious results. The subsidies led to inefficient farming practices, and the absence of market forces ultimately harmed the industry by raising food prices and putting small farmers who didn't qualify for the subsidies out of business. UBI may seem to make sense as a way to compensate for jobs lost to automation, but it simply won't work long term.

For that reason, the idea of a universal basic dividend (UBD), first proposed by economist and politician Yanis Varoufakis, has gained popularity. Supporters of UBD claim that there is a fundamental flaw in the traditional capitalistic model, where wealth is believed to be created at the individual level, while whole wealth is actually generated at the collective level. Why not redistribute some of the dividends that will be generated by public investment in AI and automation back to society? While this is an interesting mental exercise, history tends to show that capitalistic economies do not thrive from a top-down redistribution of value, due to the reduced incentive for producers to continue innovating and investing.

Without a specific vision, it would be easy for a government to be elected based on the promise of free subsidies. But UBI and UBD are not the be-all and end-all answers to the transitions we will experience in the economy. They may have utility as temporary systems to aid the transition from industrial to a digital economy. Free subsidies, however, ultimately drive public debt through the roof, leaving the public worse off long term.

These "solutions" are all predicated on the notion that AI will permanently eliminate jobs. None of them address the key issue, which is establishing an economic system where we create as many new jobs as we eliminate. They also don't consider the new types of jobs we will need to create.

AI INEQUALITY

In the eighties, the median income group's growth flattened out while the real GDP and labor productivity increased. Figure 1 shows how in the US, while real GDP has grown by 80 percent since the mid-eighties, median household income adjusted for inflation has merely gone up by 20 percent. This phenomenon doesn't seem to be happening in the US only. A recent study has observed similar outcomes in sixteen European countries.[127] The middle-income group is declining due to the number of white-collar jobs being displaced by e-commerce and the internet. Capital holders and investors are seeing increased profits while the middle class stagnates. This type of economy makes the rich richer and the poor poorer, while those in the middle are seeing themselves closer to the poorer end.

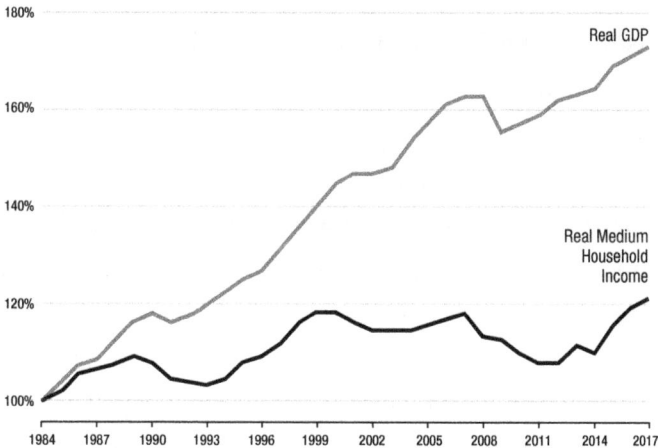

Figure 1. Real GDP Growth vs. Real Medium Household Income Growth in the US (1984–2017)

We believe this trend will continue as AI charges ahead. Intelligent automations are already infringing on white-collar jobs. More office jobs will be taken over by algorithms. Over time, we will see more blue-collar automation as well, but that requires hardware, which is more expensive. White-collar jobs don't have that hardware component, making it cheaper and easier to implement AI and see results. At the same time, those who have the capital will very likely invest their capital in technology-driven giants given the chance of making a higher return. What this will result in is more income inequality for more people, with the people at the top making much, much more money. Furthermore, with more capital being injected into technology giants, they would be better able to push for new technologies and hoard them as discussed in chapter 3.

Historian Yuval Noah Harari predicts that in the future, AI inequality will result in a group of people he terms the Useless Class. What he means by this is that there will be people who are militarily and economically useless. We will have drones to fight our wars and machines to do our work, so what good are these extra humans? It may sound a bit extreme, and people have made similar warnings for the past 200 years. But to him, at least, it is something that seems far likelier to happen this time around.[128] It's a scary thought but something we should consider when planning for the future.

IS THIS SOLVABLE?

While we like to emphasize the rosy benefits of AI and we're not entirely pessimistic about the future, we do think the repercussions on workers and income inequality need to be addressed. It's vital for governments to truly play their role in the future of AI; we need them to do three things.

First, the government needs to have a foundational understanding of what AI is. To some extent, some need to be evangelized about the importance of AI integration and its role in our future.

Second, the government needs to understand that AI should be used for specific purposes and in specific ways. We need to debunk the myths about AI's potential negative effects on the economy and encourage the government to invest in technology and infrastructure. We also need to think about the specific purposes of AI and raise questions about how to build a digital strategy as we move into the future.

Furthermore, it is necessary to give a more realistic picture of what AI and related technologies can actually do. "Policymakers don't read the scientific literature, but they do read the clickbait that goes around," said Zachary Lipton at an *MIT Technology Review*'s EmTech conference.[129] He warns that if they cannot separate hype from reality, it can lead lawmakers to put too much faith in algorithms' ability to govern things like autonomous vehicles and clinical diagnoses.[130] It is necessary to have a good understanding of the technologies and, perhaps more importantly, their limitations.

In 2017, the UK, for instance, reaffirmed a budget pledge to put driverless cars on the road by 2021.[131] This appears to be, at best, an overwhelmingly optimistic appraisal of the near future and, at worst, a complete lack of understanding of technologies. This could pose a danger to the public. For instance, a fundamental limitation of AI that doesn't get talked about often enough is "what happens if a model is trained with a set of data collected in the past but the real-world situation has changed?"[132] The model would be unequipped to handle any new or novel situations that come up. Government officials need to develop a much better understanding of how technologies work.

Lastly, the government needs to become comfortable in becoming a completely digital entity. It's important to tell governments that they have a role in the future of AI and not to discount it as just "computer stuff."

In most places in the world, people distrust the government. Unfortunately, most governments comprise people with special interests and government workers who aren't necessarily the cream of the crop. Low wages and a stigma toward public sector work has driven off some of the best talent. There isn't an atmosphere of competition that drives people to do and be their best, like in the private sector. Furthermore, the talent we see as educators doesn't tend to emphasize public service, because we haven't done a good job in promoting aspirations in public service.

This is not true everywhere. In Singapore, it's considered an honor to work in public service. In Chile, the government sends its employees to get educations abroad in places like MIT and pays their tuition. More governments need to follow these models so that they can attract the best talent and give themselves a fighting chance towards competing with, or at least conversing with, the private sector.

If we're being realistic, however, we likely won't see a lot of movement from the world's governments on these issues. Right now, most governments don't seem to want to do much to change. That's why it's important for individuals to make more noise on this issue to spread accurate information and dispel myths about AI.

So much will depend on the leaders of tomorrow, the younger generation who are more open and excited about these technologies. These individuals, some of whom are only students now, understand the situation and will hopefully employ their knowledge and run for office and create public policy that reflects a technologically bright future—both for workers and for their countries.

CHAPTER 7

HOW SHOULD WE PREPARE OUR CHILDREN FOR THE WORLD OF AI?

People are constantly asking us how we can better prepare kids to take on tomorrow's job market. This is a tough question to answer because technologies are changing at such a lightning pace that it is difficult for humans to keep up. According to one popular estimate, 65 percent of children entering primary school today will end up working in completely new jobs that don't yet exist.[133] Another source points out that as much as half of the content in a bachelor degree may be obsolete within five years.[134] Indeed, when we ourselves are struggling to stay abreast of these changes, it's difficult to definitively explain how we can prepare our children for them.

That said, there are certain things we do know about what the future will look like, and as a result, there are ways that we can prepare our kids to become more adaptive to a constantly changing environment. This reminds us of a conversation between Thomas Friedman, a renowned journalist, and Astro Teller, who oversees the Google X research and development lab.[135] They talk about how the rate of technological change has recently outpaced our ability to adapt as humans. While legislation and governing are scrambling to keep up, companies, especially technology giants, are struggling under updated regulations, all the while the public has no idea what's going on. But if we could enhance our ability to adapt even slightly, it would make a huge difference. According to the researchers, adaptability is 90 percent about optimizing for learning.

We must seek ways to learn more and faster, as well as to expose ourselves to a greater variety of issues. We will have to experiment more, try more new things. This, in turn, likely requires a different set of skills. Here, we present some ideas on what our children can do. It's not simply a matter of pushing them into the right field, or teaching them the right skills, it's about encouraging traits like resiliency and creativity that will help them deal with whatever the future holds.

REDEFINING THE CONCEPT OF A "CAREER"

When we were children, we remember vividly how our parents encouraged us to go into certain fields for professional success. They pushed us to become doctors, lawyers, or investment bankers because these jobs paid well and were guaranteed to be evergreen and always in demand. This was sometimes at odds with our passions and dreams, but back then, our parents genuinely believed they were doing the right thing to set us up for future success.

Today, however, it is difficult to forecast what career paths will guarantee lifelong success. In the future, some jobs will be around for the next five years, but we won't be able to say, "That's what you're going to do for the rest of your working life." Beyond a five-year horizon, the career landscape is hard to predict.

According to a report from the World Economic Forum (WEF), by 2030 there will be an estimated 133 million new job opportunities created due to AI, but at the same time, 75 million will be lost.[136] So we can expect to end up with a net gain of jobs in the future. However, they just won't be the same jobs as before. As physical labor jobs disappear, new positions will take their place. These are jobs that never existed before, so it's difficult to imagine exactly what they will look like.

Unlike in the past, future job prospects don't have the luxury of stable careers with guaranteed jobs. Instead, we need to redefine the word career and move away from the old idea of long-term career tracks based on specific vocations and training. In the Fourth Industrial Revolution, careers are more likely to have multiple cycles and variances.

In the past, a college degree used to guarantee you a good job and a long career. This is no longer the case. An education might still lead to employment, directly or indirectly, but it's not a guarantee. Many educated people end up pursuing jobs completely unrelated to their degrees, sometimes in fields that don't really require college degrees at all. We probably all know of an entrepreneur or two who eschewed the tech world and started their own companies offering local services like plumbing and landscaping. So while education has value, many people have been able to succeed in this new atmosphere without a traditional education.

Things have changed substantially. Our parents *found* jobs, but our children will need to *invent* their own. It's difficult to prepare children for a future where technology evolves at a faster pace than humans do. We need to prepare them for their future, not our present. Instead of encouraging our children to go into one vocation, we need to prepare them to create their own jobs.

Indeed, as parents, when we ask our children, "What do you want to be when you grow up," we are implying—and, worse yet, promising to them—this job will continue to exist in the future. Think about it: how can they dream of a career when many of the opportunities available to them in their adulthood have not yet been conceived? Instead, we are probably better off asking them, "What interests you," to find out how they can adapt to the future of work. Serial questions like "what is your passion," and "how do you make your passion productive," and "how do you direct your passion and productivity to create value" may be a much better guide to our children's futures.[137]

RETHINKING EDUCATION

Our current educational system has been around for about 250 years without any major philosophical changes. It was developed by the British in the Victorian times, which primarily wanted to train government servants and administrators to manage the then-vast empire. As a result, they created a system of specialization, where people were trained to do one thing very well.

All these years later, our educational system still functions similarly. It's quite stratified: students are asked to choose a specific discipline (or a major or a concentration) and concentrate solely on that. Academia contributes to the problem by encouraging professors to specialize very narrowly in their fields. As a result, these educators are only able to teach their specific subjects and know little about any other specialty. They are often unable to connect subjects across different disciplines.

Examination styles have also remained mostly unchanged. Education systems are exam-based and, for the most part, still reliant on memorization. In Hong Kong, children as young as three have piles of homework to complete. At age six, some kids are attempting suicide because they are so stressed by the amount of homework they have, and they often crumble under tremendous amounts of pressure. Even if they survive and are hospitalized, they are still expected to complete their assignments. This model is harmful to children and still doesn't prepare them to think critically in the real world, let alone develop the social skills needed to function well as an adult.

We use a strict binary system to measure people according to skills. We have entry tests, admission tests, intelligence quotient tests, and other various methods for trying to standardize intelligence. We have a friend whose kid is only thirteen and must face three admission tests in some of the UK's best schools to guarantee admission into the University of Cambridge or Oxford University. Our friend is a finance professor who is personally preparing his kid for a math test. This is an immense amount of pressure to put on a kid who is only thirteen years old—and the parents.

Contrast this to the model employed in Finland, which is wildly different. Children are given free time to develop and explore until the age of seven. Only then do they start their formal education. The Finnish model seems a lot more in line with preparing kids for what is to come rather than the traditional model of cramming for exams.

It is necessary to disrupt the current education model to better prepare future generations to enter the workforce. Standardized testing has pushed kids to measure themselves on an individual level, when in real life, social intelligence and team dynamics matter much more. People advance more quickly and more effectively due to their ability to build networks rather than their credentials. Those networks are social and, like an ecosystem, necessary for thriving life. But we don't teach networking in school. Instead, we keep focusing on metrics.

People who graduate from schools with pedigree and brand recognition, like Harvard Business School, do tend to get the better jobs off the bat. Yet, after five to ten years, those same graduates can be found worse off than those who graduate from more "humble" schools. Usually it is because they can't keep up with the pressure and are more likely to become depressed and have split families.

There's this story about a famous Harvard professor who went to an MBA reunion after ten years. He asked the crowd how many of them were divorced and half the class raised their hands. He then asked how many of these smart people would have raised their hands ten years ago if asked whether they'd be alienating their family ten years later.[138] It's an example of how we don't necessarily teach social intelligence. We focus on certain aspects of development but ignore the ones for coping with the pressures of life and focusing on core values like family.

As we look to the future, we need to emphasize grades and memorization skills less and emphasize the development of key soft skills more.

AN INDUSTRIAL EDUCATION MODEL HURTS KIDS

In the book *Weapons of Math Destruction*, Cathy O'Neil talks about how the introduction of algorithms into public schools has created even more inequality between wealthy and economically disadvantaged children. Often black box algorithms are being used to decide how to distribute educational funds, rather than class size or average income of the community.

When administrators are forced to be so focused on numbers like test scores, retention rates, and grades, teachers are forced to assimilate to survive. Instead of using a proper lesson plan, they start using class time teaching students how to pass exams, and sometimes even what the answers are. Kids that are outliers, the ones who are a bit more creative or entrepreneurial but not necessarily mathematically oriented or great test-takers, suffer from poor performance, metrically speaking. The principal might then call the parents and inform them their children are poor students based on exam scores alone. If the child comes from a background where the parents are disadvantaged as well, the cycle will simply continue. These kids tend to be marginalized by teachers, the adults who are supposed to care about educating children, as people who "don't do well in school," and it's largely because of standardized testing.

In contrast, many private schools aren't as dependent on standardized testing or metrics for funding purposes. As a result, test scores aren't as critical to the administration. The students also tend to have more access to private mentors who help them build skills on a more interpersonal level. The kids are given more tools to succeed. Most of these kids are already from a higher socioeconomic class, and as a result, the class divide continues to widen.

With the introduction of algorithms, governments and higher-ranked public school administrators treat education like an industry, whereas the private schools treat their students as individuals with unique challenges and skillsets. Because their educational methods and beliefs consistently pump out thoughtful, educated children, private schools can consistently raise tuition because they generate results. Public schools should take a lesson from them.

We believe in decoupling the industrial model from our educational systems. Kids are not products: you can't standardize them—they don't go through phases of production or represent specific value positions. Our standard education model is making kids weaker and, to an extent, dumber. They are unable to deal with the complexity of today's world.

Class inequality also surfaces when it comes to familiarization with technology. It's important for children to learn about new technologies sooner rather than later. Schools that are well funded can incorporate more tech into the classroom, giving students an advantage—like providing learning opportunities with laptops or iPads, for example. Schools in poorer areas can't afford the same materials, so the rich crack ahead as the poor fall more and more behind.

As discussed earlier, using standardized testing to prove that students are learning is also hamstringing educators. They are so busy teaching students how to do well on the test that they may not be helping students learn in the most effective ways. While testing makes sense for some courses, in others it simply does not. We should promote free thinking in problem-solving, not how to formulate "model answers" to complex problems.

A lot of the crises we see today are due to the training of people to think in monolithic terms. We educate people based on tenets and foundations that are almost obsolete, and that is what makes them fragile. A lot of the customs and practices in education systems are still training us on doing parts of jobs that are going to be taken over by machines. Remember, we are not losing jobs to technology, we are redefining them to increase efficiency. Technology is not to be blamed. Technology does what it's supposed to do. People don't consider that the real issue is that many jobs are poorly designed to begin with and don't take advantage of human ingenuity. We need to pivot and switch educational focus on the parts of jobs that machines cannot take.

At the same time, the organization of businesses is also changing. A job, otherwise known as engaging individuals as employees, was once the most efficient means to come up with plenty of goods and services at scale. In the past two centuries, "company" was effectively the container in which we mostly did all these things called "jobs." This model has worked well since the Industrial Revolution. And our learning systems have been set to prepare people for this kind of work environment—as specialized experts. Yet digitization has dispensed with the need for the approach to achieve scale, enabling many of today's most successful companies to act as just platforms where value is created *on* the platform by individual workers instead of fixed containers in which value is created.[139] Education and learning will therefore have to evolve.

RETOOLING EDUCATION: THE FOUR C'S

It all comes down to this: the stakeholders in education, namely parents, need to think about what aspects of the educational system are still depending on outdated models. Nobody is served by forcing a one-size-fits-all approach on a huge number of people who may learn differently. It simply encourages conformity to get ahead: whoever fits the model best advances. Those that don't conform end up falling through the cracks.

However, many parents have bought into this system and continue to perpetuate its tenets. They often demand more tutorials and more homework for their children, thinking they are giving them an edge. This is crazy: the idea that we need to give kids as much work as possible for them to perform the way they need to perform in the current educational system.

We need to eschew these outdated models and retool the

education system so that it truly trains kids for the future. If the job market has changed so much between when we were kids and the way it is today, how different will it be for our kids several decades down the line? One thing we know is that a kid's test performance in third grade math is probably not going to matter in the long run. The era of safe havens and guarantees for success in education are over.

We need to give our kids the tools they need to cope with the challenges of life. We believe we can do this by emphasizing the four Cs: creativity, coding, communication, and confidence. These are subjects that parents can prioritize helping their kids learn and excel in.

Creativity taps into the entrepreneurial spirit. Coding allows kids to start thinking about programming and the way humans and machines interact. Communication goes into their ability to express opinions both written and verbally, allowing them to improvise and speak publicly with ease. And confidence engages with their playfulness and the willingness to take risks.

CREATIVITY

Creativity is the keystone of entrepreneurship. Unfortunately, this isn't something parents mobilize their children to engage in very often. The exception is in the

US, where there is the classic image of kids creating their own lemonade stands to make a little bit of extra money. Outside of the US, though, entrepreneurship is very much a gray area. In fact, in many cultures, doing your own things and running your own businesses instead of working for a corporation is looked down upon.

Even though lemonade stands may be fun, they aren't enough. We would like to see kids learning to think creatively and take initiative through their school curriculums. They need to be encouraged to come up with creative solutions and learn how to turn ideas into actionable things. They should be able to feel the conversion rate between having an idea and turning that idea into a specific prototype.

CREATIVITY AND THE FUTURE JOB MARKET

Part of the reason this will be so important is that as AI takes over so many jobs, we will inevitably see a rise in what we call the Gig Economy. We will see more and more versatile freelancers creating their own jobs by doing piecemeal work for others.

The government and big companies are actually going to encourage this change because it benefits them financially. Some big software companies are already moving toward hiring more contract workers instead of full-time employees, because it gets them out of paying for pensions. Unfortunately, the gig economy puts workers at a disadvantage right now, because it makes it more difficult for them to save for retirement or even take out a mortgage. But as we move further ahead in a market economy, we must be prepared for this kind of reality, as new business models will likely continue to be introduced to build up these types of gig-based jobs. This trend is most likely going to continue as more and more companies are becoming platform-based rather than following the conventional command-and-control structure.

At the same time, we are seeing more and more people who work multiple jobs for different reasons than before. In the past, people picked up extra jobs because they needed extra income, but now we are seeing a transformation where people in the middle class divide their time between multiple jobs or even multiple professions not because they're underprivileged but because they're taking advantage of as many opportunities as they can. In the past, someone might solely work as a professor. Nowadays someone might be an educator, a public speaker, a company co-founder, a consultant, and an author, all at the same time.

We are also seeing the rise of the so-called *free-stack freelancers*—people who work for multiple companies doing several different things. The origin of "free-stack" comes from free-stack engineers; the term describes people who have enough knowledge to deal with every layer of software development. They can basically fit into any part of the development process, making them very versatile to take on different jobs. They don't need to be experts in one area; instead they just need to know enough to complete the job. Free-stack freelancers are those who hold a portfolio of different skills and activities instead of swearing allegiance to one employer.

We often tell our students that they will all be consultants in the future, without working for a single consulting outfit. They will most likely have multiple career tracks and knowledge bases, not just one stratified employment track. They will need that innate creativity to invent and explore these new jobs, and to recognize challenges that can be turned into job opportunities.

TECHNOLOGY HAMPERS CREATIVITY

Unfortunately, in many ways our advancing technological world works against us when it comes to fostering creativity. Constant distractions are shortening our attention spans, and time sucks like video games, YouTube, and social media often sidetrack us from productive thinking. This has happened to many of us: we struggle to come up with new ideas, so we set aside the problem for a while and return to it later, at which point all those ideas that had formerly eluded us suddenly strike. But it's difficult to refresh our perspectives—a fundamental way to create a burst of creativity—if we are facing a nonstop onslaught of incoming messages from email, SMS, Facebook, and Instagram updates, as well as tweets.

And the result of such information overload? Our thought process narrows, making it a lot more difficult for new ideas to make their way to us; there's just no space. One study in the *Harvard Business Review* even talks about how Outlook notifications are hampering productivity because each time we are distracted by them, it takes about twenty minutes to mentally reboot and refocus back to the original task.[140]

We are now able to read short messages and digest their content without retaining any of the information. Retention rates are generally very low.

Technology might make society better, but it doesn't make our brains any better. This isn't new information; it's why many Silicon Valley executives are raising their children with as little technology as possible.[141] When using tech, we become more fatigued and less productive. The brain, by nature, is a serial processor, not a multitasker. Therefore, we need to support and develop creativity; otherwise we're going to create a generation of people with zero attention spans, no focus, and no retention ability.

FOSTERING CREATIVITY

Without the process and research that went into writing our last book, *Understanding How the Future Unfolds: Using DRIVE to Harness the Power of Today's Megatrends,* we wouldn't have arrived at this volume in your hands. Analyzing different trends, inequalities, and how technology does and will impact our lives in the future has completely shifted our focuses. We saw how technology is fundamentally changing the basis of our society. That book allowed us to unlock our own creative thinking in new ways.

One of the key components of developing creativity is being able to look outside our own little world. In a sense, social media is detrimental to this skillset, especially Facebook. In theory, the internet allows us to see the world more broadly and widely, but in practice, these platforms limit us to seeing what our friends are up to. We inadvertently self-select the types of stories we see, because we choose to surround ourselves with people who have the same opinions. This prevents us from hearing other points of view and keeping an open mind. The echo chambers of WhatsApp group chats have a similar effect on narrowing our horizons. To really have a chance in the rapidly changing future, people need to be able to connect dots that were previously loose. It's the only way to engage in new thinking and assess novel ways of solving problems.

Reading as widely as possible is a key factor in fostering creativity. A lot of people these days aren't in the habit of reading. Everyone is staring at their phones all the time, but what are they really doing? They're playing computer games, watching YouTube, or scrolling through photos. The heavy dependence on online technologies is gradually taking away our ability, or at least our patience, to read longer articles. Many of us now prefer short articles because there are so many elements competing for our time and curiosity (conveniently, many online articles inform us how long it will take to browse through the article). At the same time, we have become more selective in our reading and read at a much more superficial level, too. The problem with this is that we are denying ourselves the precious opportunity to think deeply. We miss out on that meditative act that allows us to replenish our minds and to engage more deeply with an inward flow of words, ideas, and emotions.

People should be reading on paper, not computers. This is important because our phones are training our eyes to not read long text. Everything is compact and short. Reading longer works slows us down and allows us to digest content more slowly, which allows the brain to process more ideas and new perspectives. This is one reason that ebooks haven't completely eclipsed paper books; real books are both enjoyable and good for your brain.

Retention rates are also higher from reading printed books. In Norway, a study of tenth-graders showed that students that read from books scored higher on questions after reading than students who read the same text on computers.[142] The advantage isn't just in longer works; it's the actual physical aspects of books that help us retain our learning better. We can flip pages back and forth; we can also use photographic memory to help us remember things (sometimes called spatiotemporal markers). Reading on a screen, everything looks static and the same as you scroll further and further down a page. Microblogging and news retention rates, as you can imagine, are about as high as remembering a hotel room number. You forget it immediately.

Another interesting example of using creativity without technology is the concept of an escape room. This is a popular business model right now and a fun game. You are set up in a room, and you have one hour to find the necessary clues to escape it. You are not allowed to use your phone. This really throws people, because for one hour they must problem-solve, and they can't use their phones to search for information. We've trained our brain to use shortcuts because that's what technology exploits. Once the idea that there are no smartphones around sets in, then people start to look at the environment surrounding them in a different way. Then, and only then, can they escape the room.

Going low tech is a good way to increase attention and levels of creativity. This is a concept that could be emulated and employed more in an academic setting. While it is important for students to learn how to harness technology, it's equally important for students to tap into their innate creativity to problem solve.

CODING

Coding is the process used to program the binary algorithms that we use in modern tech and will likely continue to use for specific functions for the foreseeable future. Many parents believe that becoming a computer programmer will be a surefire career, just as doctors and lawyers were previously. This is not necessarily true. Not everyone is going to be a programmer in the future, but having a working understanding of what coding is and how it works will be necessary for everyone.

All three of us are strong proponents of learning a foreign language, as it is, we believe, an exceptionally useful skill for children. In the old times, it used to be classic languages, with French as the elite language. Later, it became Latin for many Catholic schools and ancient Greek in others. Now, any language will do. Among the three of us, we can claim to have the basic knowledge of at least ten languages. But the value of languages is not just improved communications; it is also about how our brains are wired as a result. We like to think that we can process our thoughts more multidimensionally. Coding is essentially like learning a foreign language. It allows us to decipher an entirely different world, a language for interacting with the world of computers.

Not too long ago, there were just a few programming courses, and only a handful of people were interested in that line of work. Then computers became a general-purpose technology, and most of the work was already done for consumers. We didn't have to worry about the programming languages anymore; we just had to enter a website address.

Coding gives kids the opportunity to start interacting and interfacing with machines in a more constructive, interactive way, which positions them for a future where machines will play a large part of our everyday lives.

IMPROVING OUR RELATIONSHIP WITH MACHINES

The real benefit to getting kids involved with coding is not just that it gives them access to more career opportunities. It's that through coding, the coder learns that the machine does what the coder tells it to do. The input defines that output. You could argue that the output might discover a relationship that you were not expecting, but it's because you have programmed or provided input in the first place that those correlations are sound.

Teaching children to code shows them that the machine is really just a means to an end, and that end is defined by the programmer. This is a way to reconcile the tension between machines and humans. It empowers them to view AI and technology as a tool for improving human quality of life and not something unknown or dangerous.

CODING AS CURRICULUM

Coding is only going to grow as a profession and as a part of our world. We would like to see learning to code held in the same academic regard as learning a foreign language. It can't be taught quite the same way, because coding isn't about beauty or aesthetics; it's, as mentioned earlier, about heuristics and more like learning spoken languages. Learning the basics of coding could play the linguistic role of helping to decipher computer communications.

Not all kids are going to be interested in learning to code. Generation Z has already made it clear that it doesn't believe that traditional learning methods should be forced on everyone, nor does it think that college is the only way to get ahead in the world. The truth is that knowing everything about coding isn't going to guarantee anyone a job or a happy life. Regardless, everyone should become familiar with coding, no matter their age, because it allows you to see what exactly coding can do for you and how it impacts your life.

Child-friendly coding programs like Scratch, Code. org, and CodeAcademy are becoming more and more common. They're designed with fun exercises with objectives so that children feel like they are playing a game and not taking a class. See Figure 1 for an example.

Figure 1. Screenshot of Scratch, a website to teach kids to code

Parents in China have taken an even more proactive approach for their children. In the absence of coding lessons in school, they pay for private lessons.[143] We would like to see more engaged coding programs incorporated into schools.

Estonia is a model example for integrating computer science into curriculum. Computer programming and IT skills are taught beginning in kindergarten. In just a few years, students are able to code games independently. These computer science classes are required; once the students finish high school, there are additional opportunities to build tech skills and to intern at tech companies.[144] Many other European countries, including Austria, Bulgaria, the Czech Republic, Poland, Lithuania, the UK, and others, have also incorporated coding into their primary and secondary school curriculums.[145]

Even if you aren't in primary or secondary school anymore, coding is still worth learning. We recommend all our business students experiment with Python because we believe it is important for them to know the basics of coding (hopefully we won't even have to do this ten years from now because all our students would have been taught some coding in secondary school). We recommend kids start with ScratchJr (scratchjr.org) to pique their interest and get them early exposure.

Consider learning to code with your child. Learning to code can be fun and potentially a bonding experience. One of the co-founders of Nexus FrontierTech lives in

Bangkok, and though his family lives in Tokyo, he spends his evenings coding with his son. They share a computer screen, and it has become a bonding experience for them. This is a great (undoubtedly geekish) example of a way to give your kid early and effective exposure in this new language.

COMMUNICATION

To be blunt: social media is killing everybody's communication skills. You can see this in how shoddily most people compose emails. They are so used to short sentences and being able to avoid face-to-face conversation that they are unable to effectively convey their thoughts. Social media is not as social as we believe it to be. Despite the name, these platforms don't necessarily improve our social skills. On the contrary, a recent study has shown that Facebook may undermine social connections.[146] Relying too much on texting—and consequently forgoing other communication methods—can gradually chip away at our ability to create and maintain social bonds. Brevity makes it more difficult for us to communicate. In a study Twitter conducted, it discovered that those who used the longer tweets got more followers and more engagement and spent more time on the site.[147]

Email, texting, and messaging apps have become a convenient way to hide from confrontation. In Japan, there is even a company called Exit that will quit your job on your behalf, because so many people are afraid to have an uncomfortable conversation in person.[148]

One way to give our kids a greater advantage in life is to teach them to be eloquent and to engage more with people away from handheld devices. A big part of communication is through nonverbal cues. The ability to make eye contact, find common ground, and actively listen is as important as being able to present ideas clearly and concisely in a professional manner. Spending time in a social setting will help youth become more at ease around other people and help them develop the communication skills they will need in the future, be it as an employee trying to get a promotion or as a freelancer who is trying to get a new consulting gig. And in the end, better communication skills lead to more empathy and more developed social intelligence.

COMMUNICATION FOR ALL

If we wish to have a society that is more socially empathetic and kind, then communication skills need to be taught to every child. Unfortunately, we see economic inequality driving an ever-increasing wedge here in what we call "the 30 million-word gap."[149] Children who grow up in low-income families are exposed to only 600 different words a day, while children in high-income households are exposed to 2,000 words a day. That means it only takes three years for a child to hear 30 million fewer words than their wealthier counterparts. Since 85 percent of a child's neural connections are formed by age three, it's too late to change things by the time a child is enrolled in school.[150] By then (around five years old), this simple vocabulary disparity is already impacting their ability to use speech as a communication tool.

Communities can get involved in bridging this gap by creating programs that make it possible for underprivileged children to be around others to talk and listen. In the US state of Georgia, the health department uses government funding to support these types of programs because it sees it as a public health issue. It's easy to see why when we look at some of the statistics. Children who can't use as many words don't do as well in school or later in life, either. They end up with shorter life expectancies and higher risk for high blood pressure, depression, and obesity. Though a smaller vocabulary cannot be the reason for all these differences between the two groups, improving a child's word bank doesn't just improve the number of words they know; it also improves their chances to be able to express themselves and lead a more fulfilling life.

BOOSTING COMMUNICATION SKILLS

As parents, we can try to enrich our children's language

abilities, both by increasing their vocabularies in their native tongue and by exposing them to foreign languages. Kids who can speak more than one language have a huge advantage when it comes to communication. This is true even if they never achieve fluency in their language of study, because the act of learning a new language develops their active listening skills, which is also part of the communication process. So, in this case, they will still see benefits in their communication skills while also learning to appreciate that not everybody speaks using the same language they do.

Another side benefit is that learning a new language enhances a child's ability to problem solve and think creatively. When learning a new language, if a child doesn't know a word or a phrase, they will have to come up with alternative ways to say what they want. For instance, if they don't know the word for "hungry," they will have to find ways to work around it by using words they do know to get the same result, like "I want to eat something." This type of cognitive challenge is both a problem-solving and creative exercise all in one.

Improved communication skills are also linked to better emotional fitness. Communication is one of the best ways to take control of some of our most predominant emotions. People with better communication skills can better express themselves verbally and not turn to other forms of communication, like using their fists. One of our friend's kids even learned sign language before learning to speak, and it prevented tantrums because she was able to express her feelings.

COMMUNICATION AND AI

The ability of a machine to communicate or interact in a more human manner is all dependent on the person behind it programming the scripts. Take Amazon's voice assistant service Alexa, for example. It started out strictly as a call-and-response device, but parents began to voice their concern that their children were losing their manners when barking orders at the device. They were talking to Alexa in a way that would never be approved of if it were said to another human being. In response, the Alexa algorithm was reprogrammed to include a manners module titled "Magic Word." It incorporates common courtesies into Alexa's interactions with humans. With Magic Word, when a speaker adds the word "please" to the verbal request, Alexa will say, "Thanks for asking so nicely."

This new programming was not a random decision. It was the result of Amazon's working with child development experts who showed that positive reinforcement works better with kids than a negative reinforcement approach ("You won't get the answer until you say please"). In effect, knowing how to communicate effectively means we will be able to create better bots in the future.

Today, scientists at the US military's Defense Advanced Research Projects Agency (DARPA) are trying to teach AI algorithms how to be more human so that they will be more easily accepted into society. If there is anything to fear about AI, it shouldn't be that jobs will be lost to them; it should be that AI becomes *too* human.[151] Because machine learning learns only through data collected on real human beings, it would be able to pick up on bias against and fear of outsiders and people from different cultures. If it began to model that behavior towards humans, we would not be able to live peacefully around AI. This possibility is just yet another reason to instill empathy in children and teach them how to use words to express themselves.

As AI starts to dominate the job market, it's more important than ever for humans to develop and strengthen the unique skills they have that make them humans. Machines are very good at processing and analyzing. They are even able to present information eloquently. But communication skills—the ability to express ideas as words—are humanity's purview. To the extent that machines can communicate well, it's only due to the person behind it programming the scripts.

CONFIDENCE

Finally, it is essential to teach our children confidence. Confidence in how people carry themselves and deal with uncertainty is a huge advantage. Confidence in saying you don't know the answer, but you will find it out. Many people fear the future because they fear instability and they fear failure. Cultivating confidence will allow our children to go boldly into the future with the agency to decide their own fate. It will help guide them toward what they want for themselves. By the time they are adults, we expect that they will not have all the answers, but if they believe they have the endurance to prosper no matter what happens, that makes all the difference.

In the future, we expect that our children will be inventing their own jobs and making their ways in an open entrepreneurship system. To prepare for this, they need to learn to take risks, fail, get back up, and move on to the next idea.

We tend to stigmatize failure rather than encourage people to go back to the drawing board and try again. Look at the way the education system demonizes mistakes by docking grades and penalizing students. It's sometimes framed as constructive feedback, but the message is essentially negative. As a result, we often run into students trying to "negotiate" for better grades rather than trying to learn from their mistakes. The current system is sending a negative message about the nature of mistakes, which is hurting kids' innate instincts to try new things. It also encourages them to blame the system instead of looking inward.

We need to shift the paradigm to encourage kids to make more mistakes. Let them know that it's fine, it's natural, and expected. In the real world, making mistakes is how you get closer to doing what you're shooting for.

BUILDING CONFIDENCE THROUGH RISK-TAKING

As parents, we sometimes try to promote safety over risk-taking behavior, but encouraging kids to try new things and take calculated risks helps to boost confidence. For a child, risk is not the same as it is for adults. For adults, taking a risk might involve buying an overpriced stock, but for a child, something as small as learning to ride a bicycle is riddled with fear and anxiety. How a child perceives trying new things is largely dependent on our behavior and reaction as parents. One of the best things we can do is to show our support and help them feel validated and let them know that their feelings are okay. It's also okay to give them time to step back and work up the courage to attempt something new and scary. Needing more time to process feelings of fear does not mean a child is destined to grow up without confidence.

However, giving a child space doesn't mean letting them

off the hook. It helps to give them a feeling of control. When it's time to eat something new, for example, parents can give them some form of decision-making. Let them choose how big the bite needs be or let them know they don't have to finish a portion as long as they take a bite and taste it first. And when a child does complete an attempt at something, celebrate their effort, not the result.

Lastly, another way to help kids to take risks is to model that behavior ourselves as parents. We too must be willing to try new things. As we make taking risks a norm rather than a singular, one-time event, we can help children build confidence by trying new things. We are doing our children a service by asking them to take risk.

ATTITUDE BEGETS CONFIDENCE

Confident students don't do better in school or make fewer mistakes; they're just not afraid to participate. Being willing to try something without fear of failure is what defines a strong individual—a person who won't just say, "I don't know if I can do it." If a person lacks confidence, they will not have the courage needed to pave a future for themselves or the fortitude to get through trying times for better tomorrows.

When Danny was thirteen years old, he spotted an arbitrage opportunity: buying components in his home city and building personal computers (PCs) out of them to sell to people living in countryside villages. Having put together the PCs, he would drive (yes, drive, speaking of risk-taking) to the villages where people were only beginning to hear about PCs. There was no guarantee that success would be forthcoming, but he tried. The biggest gain to Danny was not the money made from the trade. Rather, it set him on a path to keep trying new things and ideas. He was gaining self confidence in his ability to achieve goals.

Sometimes, the best way to help youths build confidence is not to help them at all. Don't correct a student if they give a wrong answer. Instead, just let them try to figure out what went wrong on their own. Given this experience enough times, the student will become accustomed to trying more than once to get something right. Even more than that, they learn to trust themselves and use their own minds to get to the right answer, or goal, or target, rather than relying on an adult.

Sometimes, of course, a student will need some help. But to build a student's confidence, it's not enough to tell them what they did wrong. You also must tell them what they did right. Maybe they remembered to apply something they learned the week before, or they tried to solve the problem in an unexpected way. Give them praise for these efforts so they learn to value the process of getting to a solution as well as the solution itself. We can also build confidence in our children by motivating them to pursue new competencies and areas of knowledge. An internal sense of motivation will be key to preparation for the future.

GOAL-SETTING AS A PART OF LEARNING

To have faith in their own abilities, youngsters need to have a sense of accomplishment. That means they should know what their goals and objectives are before they begin. Goals should be specific and realistic. They can be presented by parents or teachers or developed with the child. For instance, let's say the goal is to learn an instrument. After the goal is set, more concrete objectives, like being able to play a specific song in a certain amount of time and a plan to take lessons and practice daily, should be the next step. Measurable objectives will help children see more easily how they have advanced from being able to play nothing to playing something. Being able to recognize accomplishments is all part of building confidence.

Confidence developed in youth will carry on into adulthood. To have a successful professional life, confidence is one of the single most important character traits to possess—the reason being that it makes a person more willing to take risks, more willing to admit when they don't know something, and more willing to learn. This willingness to do more in general is what helps a person become more accomplished and recognized for their work, and overall more marketable, allowing them to take more control of their professional life when the traditional nine-to-five career no longer exists. Confidence will give them the self-reliance to build a future for themselves out of what opportunities are available to them, no matter what tomorrow's AI-driven economy may hold.

TOWARDS AN UNKNOWN FUTURE

Teaching our kids the four C's is vital in providing them with a solid foundation to face the future. These skills are not just essential; they are the bare minimum we should emphasize to help kids get a leg up in our technology-driven world.

We don't know what will happen in the next ten or even five years, but we do know today's formal education isn't going to prepare our kids for the economy of tomorrow. Focusing on creativity, coding, communication, and confidence will teach our future leaders to pivot and adapt to their surroundings, whatever they may be. With the four C's, they will be able to process information effectively, better express themselves, and confidently understand the relationship between humans and machines.

All of this will be far more valuable than a perfect school

record or an advanced degree—it will help them live a satisfactory and fulfilling life in a world filled with uncertainty.

CONCLUSION

The lack of understanding around AI is standing in our way of making progress as a society. More of the conversations around this technology need to center on empowering humans or generating human centered technology. Being a human is the one thing machines will never be able to replicate. These conversations shouldn't just be happening among scientists and academics and in think-tanks. They need to become a part of the public discourse. Right now, there is relatively little conversation in the public sphere, and as a result, the public sector is absent at the table. If the public sector doesn't step up to its rightful place, then the private sector will dominate not only technological development but the direction of society.

This is more crucial than ever as public dialogues are currently falling apart. The election of Donald Trump and the Brexit vote were two events that exposed how fragile our liberal systems really were. No amount of technology can fix a broken society. Mismanagement and misuse of technology has lent credence to extremely populist, rhetorical fearmongering around jobs being lost to machines. We need to reframe this conversation to show how technology has simply improved efficiency. Instead of eliminating jobs, AI is forcing us to rethink our standard conceptions of industry, in favor of new paradigms.

We also must initiate more meaningful discussions on how AI technologies can be detrimental to our societies and us as individuals. For instance, in classrooms over the past years, we are seeing more and more students looking at their smartphones and texting, neglecting our teaching. Admittedly, this may be a product of our being extremely boring as teachers. But it is also probable that these students no longer have the strength nor ability to fight the temptation of WhatsApp, Facebook, or even online shopping. And even when they are not gluing themselves to the screens, many would not be bothered to take notes—they would prefer to just sit there and take pictures of the board when required.[152] Past research has shown that taking notes can improve students' learning. Indeed, the quantity of notes taken is directly related to how much information can be retained by the brain.[153] The distractions have taken a toll on the inclination to learn—in an era where the willingness to learn is essentially a survival skill.

Governments also need to catch up on the greater conversation surrounding AI technology and policy, but we cannot leave it up to them. Societies, businesses, and media are the ones who have the responsibility to unpack what AI can and cannot do. With the wrong expectations, governing bodies will come up with silly or just downright dangerous policies and regulations. Driverless cars are still not ready to be deployed for many reasons, yet some governments are saying they want them on the roads by a certain date. This is a little crazy. We need to hold governments accountable for their words and encourage them to be more precise when it comes to talking about AI.

We also need to help our governments understand that AI is not an overarching panacea but a narrow, problem-solving tool. Social problems will always need to be hashed out by humans, because simple engineering solutions will never be enough to fix societal ills. If we can come up with new, outside-the box-solutions, however, AI is a wonderful tool for deploying them.

WHERE ARE WE GOING?

In the years to come, we will be forced to reckon with the ethical and practical questions surrounding AI technology. Whether we like it or not, AI is going to become a presence in our lives (and in many ways, it already is). We can't stop progress, so we need to learn how to use this technology effectively for our own good, for humanity's good.

We firmly believe that AI will create new jobs, opportunities, and infrastructures that we can't yet even conceive of. We are also afraid that without multiple stakeholders in the arena of developing AI, there will be a lack of cohesion in terms of global interest. This could be dangerous in the long run.

Whether AI is ultimately our friend, or our foe, is completely dependent on the people who conceive, engineer, and employ it. Like any technology, AI can be used for destructive purposes, but it can also become a positive creative force. As a society, we can prepare ourselves for the AI revolution through education and through redefining our relationship with technology. This includes creating a new kind of moral compass towards the development of AI. We need to bring experts into the conversation and be aware of biases that can occur in programming.

TAKING BACK THE REINS

We don't find the development of AI's existence only in

the hands of a few private corporations appealing. Unfortunately, this is where we are currently headed with the dominance of FAANG and BAT. We need to encourage our world governments to step in to give other entities in the public and private sector a fighting chance.

Since change is so rapid, the only way we can make sure we thrive, as adults or children, is by keeping up with trends and current affairs. Read the newspapers online (or better yet, the paper version) and follow the companies' news. It's important to know who is driving the push for AI, what they are creating, and what their goals are.

We also should be aware of how AI works in terms of decision-making. Imagine if we let AI make decisions on criminal convictions based solely on statistics. For example, if you're a man being tried for murder by a machine, it will judge you much less favorably simply due to the statistical nature of the input data. Statistically, three out of four murders are committed by men, and machines being fed with such data are likely to put men in a disadvantageous position. It has nothing to do with discrimination or bias. It is just historical fact.[154] We need to think about all these factors and figure out solutions to make AI more equitable.

We must get very involved in the conversations about the downsides to AI. The real danger isn't the machines but human beings themselves and the ways they program and employ AI.

Imagine there are people out there who are yearning to experience firsthand what it was like to be fighting on the Normandy beaches on D-Day as part of the allied campaign to liberate France and Europe. This could be possible through augmented and virtual reality devices. If the demand could be turned into profit, the developer would come up with a simulation to celebrate the heroic acts. Now, what happens if there also exists a bunch of individuals who want the thrill of beheading captives like some terrorist groups do? Is it ethical or moral to produce such a simulation? How do you draw the line between these two demands? As long as the latter represents a profitable business opportunity, a company is likely to assert itself to introduce a simulation. Is this right?[155]

Although this book is titled *The AI Republic*, we believe this book is less about AI and more about the human experience in the age of AI. Nothing happens in a void, and as technology evolves, it's important to use those advances as a lens to examine humanity's role in our increasing dependence on technology.

There will always be obstacles and unforeseen developments that will impede AI development. For this reason, we should seriously entertain more things like the Magna Carta for Inclusivity and Fairness in the Global AI Economy, which will project the set of rules we see as necessary for moving forward.[156]

For those who need a refresher course on the original Magna Carta, it was a groundbreaking document from 1215 that acknowledged human rights for the first time. Created in response to the abuse of power committed by King John of England, the Magna Carta guaranteed the rights of citizens, who, after the signing of the Magna Carta, could no longer be persecuted and thrown in prison without just, legal cause.

As the Fourth Industrial Revolution further shapes our world, we must continue to reaffirm our rights, this time in a high-tech, AI-driven society. If government does not get involved, we must get active ourselves. It's in our DNA. We humans did this for our individual human rights after Martin Luther separated from the church, after the Declaration of Independence, and after WWII (with the establishment of the UN). There's nothing wrong with thinking about the role that humans have in a changing landscape. Moreover, no matter what, the conversation needs to be owned by everyone.

Though the Magna Carta for Inclusivity and Fairness in the Global AI Economy has been proposed, it has not been written yet. It should be created not by three academics but by a collaborative means, developed using a diversity of perspectives from experts around the world. Our vision is an inclusive charter of rights that shapes how we develop accountability and ethically driven AI. It is our hope that such a new Magna Carta for the twenty-first century will protect not only our rights but those of our children and our children's children.

ABOUT THE AUTHORS

DR. TERENCE TSE

Dr. Terence Tse is a co-founder of Nexus FrontierTech, which customizes artificial intelligence products for its clients to build up new capabilities to attain unfair business advantage. He is also a professor at the London campus of ESCP Europe Business School.

Terence is the co-author of the bestseller *Understanding How the Future Unfolds: Using DRIVE to Harness the Power of Today's Megatrends.* Along with Mark, he was named one of the eighteen new voices in 2018 that reshape management and leadership by *Professional Manager in the UK.* Talent Quarterly in the US called the DRIVE framework one of the twenty-four trends transforming talent management in the years to come.

He also wrote the book *Corporate Finance: The Basics*.

In addition to consulting for the EU and UN, Terence has written more than one hundred articles and regularly provides commentaries on the latest current affairs, market developments, education, artificial intelligence, and blockchain in many outlets, including the *Financial Times*, *The Guardian*, *The Economist*, *CNBC*, *Les Echos*, *the World Economic Forum*, and the *Harvard Business Review* (Arabic, Chinese, English, French, Italian, and German). He has also appeared on radio and television shows on China's CCTV, Channel 2 of Greece, France 24, Japan's NHK, and Radio Romănia Cultural.

Moreover, he delivered speeches at the UN, International Monetary Funds, and International Trade Centre. Invited by the government of Latvia, he was a keynote speaker at a Heads of Government Meeting alongside the Premier of China and Prime Minister of Latvia. Terence has also been invited to speak at corporate events in Croatia, Ecuador, India, Italy, Monaco, Norway, Qatar, Russia, Thailand, the UK, and the UAE.

He has either worked with or run training for companies including Allianz, Atlantic Grupa, China Merchant Securities, Costa Crociere, Dukat, EY, F&C Investment Management, Ferrari World, France Telecom/Orange, ICICI, Indian Oil, Indian Railway, London Stock

Exchange Group, Lloyds Bank, McKinsey & Co., MOL, Molex, Monte dei Paschi di Siena, NIS/Gazprom, Olayan, Papyus, Pfizer, Podravka, Rexam, UniCredit, Walgreens Boots Alliance, Yahoo!, and YIT.

Previously, he worked in mergers and acquisitions at Schroders, Citibank, and Lazard Brothers in Montréal and New York. Terence also worked in London as a consultant at EY focusing on UK financial services. He obtained his doctoral degree from the Cambridge Judge Business School, University of Cambridge, UK.

DR. MARK ESPOSITO

Dr. Mark Esposito is a co-founder of Nexus FrontierTech, a leading global firm providing AI solutions to a variety of clients across industries, sectors, and regions. In 2016, he was listed on the Radar of Thinkers50 as one of the thirty most prominent business thinkers on the rise, globally.

Mark is a professor of Business & Economics at Hult International Business School, as well as member of the faculty at Harvard University's Division of Continuing Education, where he teaches Business, Government, and Society, as well as Economic Strategy and Competitiveness, and serves as Institutes Council Co-Leader at the Microeconomics of Competitiveness program (MOC) at the Institute of Strategy and Competitiveness at Harvard Business School.

He holds fellowships with Judge Business School in the UK, as part of the Circular Economy Center, as well as with the Mohammed Bin Rashid School of Government in Dubai.

Mark has been appointed as a global expert for the Fourth Industrial Revolution at World Economic Forum, where he contributed to five cross-disciplinary reports and worked with the Forum on Accelerating the Circular Economy as well as in the Future of Production. His next project with the WEF will be on Globalization 4.0 and the global architecture of policy in the new nascent world order.

He is a prolific author, and his articles can be found on ResearchGate. He has authored and co-authored eleven books.

Mark is the co-author of the bestseller *Understanding How the Future Unfolds: Using DRIVE to Harness the Power of Today's Megatrends*. The framework contained therein was nominated for the CK Prahalad Breakthrough Idea Award by Thinkers50, the most prestigious award in business thought leadership. The DRIVE framework has also led Chartered Management Institute's own magazine,

Professional Manager in the UK, to name Mark as one of the eighteen new voices in 2018 that reshape management and leadership. He obtained his PhD from the International School of Management in Paris/New York and his second doctoral degree from Ecole des Ponts Paris Tech, one of France's oldest and most prestigious Grande Ecoles.

He is fluent in six languages, and in his nonacademic life, he holds the rank of Master Instructor, Black Belt IV Dan, at the International Budo Institute in Canada.

DANNY GOH

Danny is a serial entrepreneur and an early-stage investor. He is the co-founder and CEO of Nexus FrontierTech, an AI research firm that easily integrates AI into organizations' processes by using natural language processing to transform idle information into structured data, enabling them to run better, leaner, and faster.

Danny is the General Partner of the G&H Ventures fund, which invests in early-stage startups primarily in Southeast Asia. G&H Ventures has invested in more than twenty portfolios in deep tech and is building its second and third fund to help startups into the growth stage.

He has also co-founded Innovatube Frontier Labs

(IFL), which later merged with Nexus. IFL is a technology group that operates an R&D lab in software and AI developments and acts as an incubator to foster the local startup community in Southeast Asia. IFL has a team of researchers and engineers to develop cutting-edge deep technology to help startups and enterprises bolster their operational capabilities.

Danny currently serves as an Entrepreneurship Expert at the Said Business School, University of Oxford, and is also an appointed Fellow at the Center for Policy and Competitiveness at the École des Ponts Business School. He is an advisor and judge to several technology startups and accelerators including Microsoft Accelerator, Startupbootcamp IoT, and LBS Launchpad. Danny serves as a visiting lecturer at various universities in Europe, and he is a speaker at various conferences, including TEDx and Fintech events.

Danny has lived on four different continents in the last twenty years—in Sydney, Kuala Lumpur, Boston, and London—and constantly finds himself traveling.

NOTES

1   *New York Times* 'Did A Human Or A Computer Write This?', Sunday Review, Quiz, 7 Mar. 2015 https://www.nytimes.com/interactive/2015/03/08/opinion/sunday/algorithm-human-quiz.html (accessed on 1 October 2018)

2   https://www.nextrembrandt.com

3   We have decided not to include the actual picture of *The Next Rembrandt* here because we cannot tell to whom the copyrights and authorship are due. See Lanza, Emily 'Who Painted Rembrandt? Copyright And Authorship Of Two Rembrandt Portraits', *The Legal Palette*, 21 Feb. 2018 for implications <https://www.thelegalpalette.com/home/2018/2/21/who-painted-rembrandt-copyright-and-authorship-of-two-rembrandt-portraits> (accessed on 26 December 2018)

4   Kharpal, Arjun 'A.I. Will Be 'Billions Of Times' Smarter Than Humans And Man Needs To Merge With It, Expert Says' 13 Feb. 2018 <https://www.cnbc.com/2018/02/13/a-i-will-be-billions-of-times-smarter-than-humans-man-and-machine-need-to-merge.html> (accessed on 11 October 2018)

5   Dobbs, Richard, Manyika, James and Woetzel, Jonathan *No Ordinary Disruption: The Four Global Forces Breaking All the Trends*, (New York: PublicAffairs), 2016

6   Statista. 'Number of monthly active WeChat users from 3rd quarter 2011 to 3rd quarter 2018 (in millions)', 2018 <https://www.statista.com/statistics/255778/number-of-active-wechat-messenger-accounts> (accessed on 26 November 2018)

7   European Commission *Communication From The Commission To The European Parliament, The European Council, The Council, The European Economic And Social Committee And The Committee of the Regions: Artificial Intelligence For Europe*, 25 Apr. 2018; Berggruen, Nicolas and Gardels, Nathan 'A Wakeup Call For Europe', *The WorldPost*, 27 September 2018 <https://www.washingtonpost.com/news/theworldpost/wp/2018/09/27/europe/?utm_term=.e984a5543bfa> (accessed on 26 November 2018)

8   Andreessen, Marc 'Why Software Is Eating The World', *The Wall Street Journal,* 20 Aug. 2011 <https://www.wsj.com/articles/SB10001424053111903480904576512250915629460> (accessed on 26 December 2018)

9   Ip, Greg 'The Antitrust Case Against Facebook, Google and Amazon', *The Wall Street Journal,* 16 Jan. 2018 <https://www.wsj.com/articles/the-antitrust-case-against-facebook-google-amazon-and-apple-1516121561>; Brandom, Russell 'The Monopoly-Busting Case Against Google, Amazon, Uber, And Facebook,' *The Verge,* 5 September 2018, <https://www.theverge.com/2018/9/5/17805162/monopoly-antitrust-regulation-google-amazon-uber-facebook> (both accessed on 25 November 2018)

10 Aley, Doug 'It's Hard to Compete With Tech Giants Like Google and Amazon— But It Can Be Done', Entrepreneur Europe, 18 Jul. 2018 <https://www.entrepreneur.com/article/316376> (accessed on 25 November 2018)

11 Kahnemann, Daniel *Thinking Fast And Slow,* (London: Penguin), 2011

12 *Exponential Wisdom Podcast,* Episode 33, 31 March 2017, <http://podcast.diamandis.com/2017/03/31/episode-33-evolution-of-humanity-with-ai/> (accessed on 27 December 2018)

13 Huang, Eustance '"Never Underestimate Human Stupidity," Says Historian Whose Fans Include Bill Gates And Barack Obama', CNBC.com, 15 Jul. 2018 <https://www.cnbc.com/2018/07/13/never-underestimate-human-stupidity-says-historian-and-author.html> (accessed on 18 December 2018)

14 Susskind, Jamie *Future Politics: Living Together in a World Transformed by Tech,* (Oxford: Oxford University Press), 2018

15 Tufekci, Zeynep *Twitter and Tear Gas: The Power and Fragility of Networked Protest,* (US: Yale University Press, 2018)

16 Oxford Dictionary. <https://en.oxforddictionaries.com/definition/republic> (accessed on 27 November 2018)

17 Griffin, Andrew 'Facebook's Artificial Intelligence Robots Shut Down After They Start Talking To Each Other In Their Own Language' *Independent,* 31 July 2017 <https://www.independent.co.uk/life-style/gadgets-and-tech/news/facebook-artificial-intelligence-ai-chatbot-new-language-research-openai-google-a7869706.html> (accessed on 28 November 2018)

18 Samonite, Tom 'No, Facebook's Chatbots Will Not Take Over the World', *Wired.com,* 1 August 2017 <https://www.wired.com/story/facebooks-chatbots-will-not-take-over-the-world/>; Emery, David 'Did Facebook Shut Down an AI Experiment Because Chatbots Developed Their Own Language?' Snopes, 1 August 2017 <https://www.snopes.com/fact-check/facebook-ai-developed-own-language/> (accessed on 28 November 2018)

19   Kucera, Roman 'The Truth Behind Facebook AI Inventing A New Language,' *Towards Data Science*, 7 August 2017 <https://towardsdatascience. com/the-truth-behind-facebook-ai-inventing-a-new-language-37c5d680e5a7> (accessed on 28 November 2018)

20   Cadwalladr, Carole 'Are The Robots About to Rise? Google's New Director Of Engineering Thinks So...', *The Guardian*, 22 February 2014 <https://www. theguardian.com/technology/2014/feb/22/robots-google-ray-kurzweil-terminator-singularity-artificial-intelligence> (accessed on 28 November 2018)

21   Searle, John R. 'Minds, Brains, and Programs,' *Behavioral and Brain Sciences*, 1980, 3 (3), 417–457

22   Gibbs, Samuel 'Elon Musk: Artificial Intelligence Is Our Biggest Existential Threat', *The Guardian*, 27 October 2014 https://www.theguardian.com/ technology/2014/oct/27/elon-musk-artificial-intelligence-ai-biggest-existential-threat (accessed on 28 November 2018)

23   Rawlinson, Kevin 'Microsoft's Bill Gates insists AI is a threat', BBC, 29 January 2015 <https://www.bbc.com/news/31047780> (accessed on 28 November 2018)

24   Cellan-Jones, Rory 'Stephen Hawking Warns Artificial Intelligence Could End Mankind', BBC, 2 December 2014 <https://www.bbc.com/news/ technology-30290540> (accessed on 28 November 2018)

25   Browne, Ryan 'Elon Musk says global race for A.I. will be the most likely cause of World War III', CNBC, 4 September 2017 <https://www.cnbc.com/2017/09/04/ elon-musk-says-global-race-for-ai-will-be-most-likely-cause-of-ww3.html> (accessed on 29 November 2018)

26   Clifford, Catherine 'Elon Musk: 'Robots Will be Able To Do Everything Better Than Us'', CNBC, 17 July 2017 <https://www.cnbc.com/2017/07/17/elon-musk-robots-will-be-able-to-do-everything-better-than-us.html> (accessed on 29 November 2018)

27   Clifford, Catherine 'Bill Gates: I Do Not agree with Elon Musk about A.I. 'We shouldn't panic about it'', CNBC, 25 September 2017 <https://www.cnbc. com/2017/09/25/bill-gates-disagrees-with-elon-musk-we-shouldnt-panic-about-a-i.html> (accessed on 28 November 2018)

28   Searle, John R. 'Minds, Brains, and Programs,' *Behavioral and Brain Sciences*, 1980, 3 (3), 417–457

29   Friedman, Thomas L. *Thank You for Being Late: An Optimist's Guide to Thriving in the Age of Accelerations: Pausing to Reflect on the Twenty-First Century*, (London: Penguin, 2017)

30 Zittrain, Jonathan 'Engineering An Election', *Harvard Law Review Forum*, 20 June 2014 <https://harvardlawreview.org/2014/06/engineering-an-election/> (accessed on 29 November 2018)

31 O'Neil, Cathy, *Weapons of Math Destruction*, (New York, Random House, 2016)

32 Chui, Michael, Manyika, James, Miremadi, Mehdi 'How Many of Your Daily Tasks Could Be Automated?' *Harvard Business Review*, 14 December 2015 https://hbr.org/2015/12/how-many-of-your-daily-tasks-could-be-automated (accessed on 29 November 2018)

33 Tse, Terence and Esposito, Mark *Understanding How the Future Unfolds: Using DRIVE to Harness the Power of Today's Megatrends* (UK: Lioncrest, 2017)

34 Frey, Benedikt Carl and Osborne, Michael *The Future of Employment: How Susceptible Are Jobs To Computerisation?*, September 2013

35 Nedelkoska, Ljubica and Quintini, Glenda 'Automation, Skills Use And Training', *OECD Social, Employment and Migration Working Papers*, No. 202, OECD Publishing

36 Bessen, James 'How Computer Auotmation Affects Occupations: Technology, Job, And Skills,' *Boston University School of Law & Economics Working Paper No. 15-49*, 2016

37 Tett, Gillian. 'How Robots Are Making Humans Indispensable', *Financial Times*, 22 December 2016 <https://www.ft.com/content/da95cb2c-c6ca-11e6-8f29-9445cac8966f> (accessed on 29 November 2018)

38 Tse, Terence and Esposito, Mark *Understanding How the Future Unfolds: Using DRIVE to Harness the Power of Today's Megatrends* (UK: Lioncrest, 2017)

39 Kremer, Michael 'The O-Ring Theory of Economic Development', *Quarterly Journal of Economic*, 1993, 108 (3), 551-575

40 Autor, David 'Why Are There Still So Many Jobs? The History and Future of Workplace Automation', *Journal of Economic Perspectives*, 2015, 29(3), 3-30

41 Shestakofsky, Benjamin 'Working Algorithms: Software Automation and the Future of Work', *Work and Occupations*, 2017, 44(4), 376-423

42 Bessen, James *Learning by Doing: The Real Connection between Innovation, Wages, and Wealth* (US: Yale University Press, 2015)

43 "Human Need Not Apply", YouTube Video https://www.youtube.com/watch?v=7Pq-S557XQU (accessed on Jan 22, 2019)

44 Dickerson, Desmond 'No Hands: The Autonomous Future of Trucking', *Center For The Future Of Work*, September 2018 <https://www.cognizant.com/whitepapers/no-hands-the-autonomous-future-of-trucking-codex3867.pdf> (accessed on 29 November 2018)

45 Williamson, Jeffrey, Did British Capitalism Breed Inequality? (Boston: Allen and Unwin), 1985

46 Schwab, Klaus 'The Fourth Industrial Revolution: what it means, how to respond', WEF, Jan 2016 <https://www.weforum.org/agenda/2016/01/the-fourth-industrial-revolution-what-it-means-and-how-to-respond/> (accessed on 19 January 2019)

47 Cross, Tim 'After Moore's Law', The Economist, 12 Mar 2016 <https://www.economist.com/technology-quarterly/2016-03-12/after-moores-law> (accessed on January 18, 2019)

48 World Economic Forum, 'Technology and innovation for the future of production: accelerating value creation'. Mar 2017 <http://www3.weforum.org/docs/WEF_White_Paper_Technology_Innovation_Future_of_Production_2017.pdf> (accessed on January 19, 2019)

49 Singh, Tarry, 'Why Data Scientists Are Crucial For AI Transformation,' Forbes, 13 Sep 2018 <https://www.forbes.com/sites/cognitiveworld/2018/09/13/why-data-scientists-are-crucial-for-ai-transformation/#65df7eb83f6f> (accessed on January 18, 2019)

50 Nsengimana, Jean Phibert. 'How Africa Wins the Fourth Industrial Revolution', Forbes, 10 Oct 2018 <https://www.forbes.com/sites/startupnationcentral/2018/10/10/how-africa-wins-the-4th-industrial-revolution/#71826e92f371 (accessed on January 19, 2019)

51 Cowen, Tyler, 'Computing the Social Value of Uber (It's High)', Bloomberg, 8 Sep 2016 <https://www.bloomberg.com/opinion/articles/2016-09-08/computing-the-social-value-of-uber-it-s-high> (accessed on January 16, 2019)

52 Klinges, David, 'A new dimension to marine restoration: 3D printing coral reefs', Mongabay.com, 27 Aug 2018 <https://news.mongabay.com/2018/08/a-new-dimension-to-marine-restoration-3d-printing-coral-reefs/> (accessed on January 18, 2019)

53 Netburn, Deborah, 'Newly discovered bacteria can eat plastic bottles', 11 March 2016 <http://phys.org/news/2016-03-newly-bacteria-plastic-bottles.html> (accessed on January 19, 2019)

54 Stanford News, 'Stanford scientists make renewable plastic from carbon dioxide and plants', Mar 2016 <http://news.stanford.edu/news/2016/march/low-carbon-bioplastic-030916.html> (accessed on January 18, 2019)

55   Shahrzad, Darafsheh 'How Does Brain Sensing Technology Work?' iotforall.com, 13 September 2017 <https://www.iotforall.com/brain-sensing-technology-muse-headband/> (accessed on January 18, 2019)

56   Choi, Charles Q., 'AI Can Help Patients Recover Ability to Stand and Walk', IEEE Spectrum, 19 Jul 2017 <https://spectrum.ieee.org/the-human-os/biomedical/devices/ai-can-help-patients-recover-ability-to-stand-and-walk> (accessed on 18 January 2019)

57   Bhatia, Richa. 'Understanding Moravec's Paradox And Its Impact On Current State Of AI', Analytics India Magazine, 7 Jun 2018 <https://www.analyticsindiamag.com/understanding-moravecs-paradox-and-its-impact-on-current-state-of-ai/> (accessed on January 18, 2019)

58   Sanburn, Josh, 'How Smart Traffic Lights Could Transform Your Commute' Time, 5 May 2015<http://time.com/3845445/commuting-times-adaptive-traffic-lights/> (accessed on 18 January 2019)

59   This section draws heavily from the discussion appeared in McAfee, Andrew and Brynjolfsson, Erik *Machine, Platform, Crowd: Harnessing Our Digital Future*, (New York: W. W. Norton & Company, 2017)

60   Hutchins, John '"The whisky was invisible", or Persistent myths of MT', *MT News International*, June 1995

61   Polanyi, Michael *Personal Knowledge* (Chicago: University of Chicago Press, 1958)

62   Autor, David 'Why Are There Still So Many Jobs? The History and Future of Workplace Automation', *Journal of Economic Perspectives*, 2015, 29(3), 3–30

63   Agrawal, Ajay, Gans, Joshua, and Goldfarb, Avi *Prediction Machines: The Simple Economics of Artificial Intelligence* (Boston: Harvard Business School Press, 2018)

64   Kelly III, James E. Computing, Cognition And The Future Of Knowing: How Humans And Machines Are Forging A New Age Of Understanding, 2015 <https://www.digintel.net/wp-content/uploads/2017/11/Computing_Cognition_WhitePaper.pdf> (accessed on 1 December 2018)

65   Puget, Jean Francois 'What Is Machine Learning?' IBM Community, May 18 2016 <https://www.ibm.com/developerworks/community/blogs/jfp/entry/What_Is_Machine_Learning?lang=en_us> (accessed on 1 December 2018)

66   Garychl 'Applications of Reinforcement Learning in Real World', *Towards Data Science*, 2 August 2018 <https://towardsdatascience.com/applications-of-reinforcement-learning-in-real-world-1a94955bcd12> (accessed on 1 December 2018)

67 https://www.youtube.com/watch?v=V1eYniJoRnk&vl=en&ytbChannel=null (accessed on 1 December 2018)

68 Fry, Hannah *Hello World: How to Be Human in the Age of the Machine*, (London: Doubleday, 2018)

69 Liang, Xiaoyuan, Du, Xusheng, Wang, Guiling and Han, Zhu 'Deep Reinforcement Learning For Traffic Light Control In Vehicular Networks', *IEEE Transactions on Vehicular Technology*, 20(20), 2018; Zhao, Xiangyu, Xia, Long, Zhang, Liang, Ding, Zhuoye, Yin, Dawei Yin, Tang, Jiliang 'Deep Reinforcement Learning For Page-Wise Recommendations' *RecSys*, 2018 https://arxiv.org/pdf/1805.02343.pdf (accessed on 1 December 2018)

70 *The Economist* 'From Not working To Neural Networking', 25 June 2016

71 Ibid.

72 Meeker, Mary 'Internet Trends Report 2014', *KPCB*, 28 May 2014 <https://www.slideshare.net/kleinerperkins/internet-trends-2014-05-28-14-pdf> (accessed on 2 December 2018)

73 Eveleth, Rose 'How Many Photographs of You Are Out There In the World?', *The Atlantic*, 2 November 2015 <https://www.theatlantic.com/technology/archive/2015/11/how-many-photographs-of-you-are-out-there-in-the-world/413389/> (accessed on 2 December 2018)

74 McAfee, Andrew and Brynjolfsson, Erik *Machine, Platform, Crowd: Harnessing Our Digital Future*, (New York: W. W. Norton & Company, 2017)

75 McAfee, Andrew and Brynjolfsson, Erik *Machine, Platform, Crowd: Harnessing Our Digital Future*, (New York: W. W. Norton & Company, 2017)

76 Executive Office Of The President *Big Data: A Report On Algorithmic Systems, Opportunity, And Civil Rights*, May 2016 <https://obamawhitehouse.archives.gov/sites/default/files/microsites/ostp/2016_0504_data_discrimination.pdf> (accessed on 2 December 2018)

77 Tucker, Ian '"A White Mask Worked Better": Why Algorithms Are Not Colour Blind', *The Guardian*, 28 May 2017 <https://www.theguardian.com/technology/2017/may/28/joy-buolamwini-when-algorithms-are-racist-facial-recognition-bias> (accessed on 2 December 2018)

78 Titcomb, Richard 'Robot Passport Checker Rejects Asian Man's Photo For Having His Eyes Closed', *The Telegraph*, 7 December 2016 <https://www.telegraph.co.uk/technology/2016/12/07/robot-passport-checker-rejects-asian-mans-photo-having-eyes/> (accessed on 2 December 2018)

79   Barr, Alistair 'Google Mistakenly Tags Black People as "Gorillas," Showing Limits of Algorithms', *The Wall Street Journal*, 1 July 2015 <https://blogs.wsj.com/digits/2015/07/01/google-mistakenly-tags-black-people-as-gorillas-showing-limits-of-algorithms/> (accessed on 2 December 2018)

80   Hern, Alex 'Flickr Faces Complaints Over 'Offensive' Auto-Tagging For Photos', *The Guardian*, 20 May 2015 <https://www.theguardian.com/technology/2015/may/20/flickr-complaints-offensive-auto-tagging-photos> (accessed on 2 December 2018)

81   Zhao, Jieyu, Wang, Tianlu, Yatskar, Mark, Ordonez, Vicente and Chang, Kai-Wei 'Men Also Like Shopping: Reducing Gender Bias Amplification using Corpus-level Constraints', *Proceedings of the 2017 Conference on Empirical Methods in Natural Language Processing*, September 2017

82   Susskind, Jamie *Future Politics: Living Together in a World Transformed by Tech*, (Oxford: Oxford University Press), 2018

83   Eubanks, Virgina *Automating Inequality: How High-Tech Tools Profile, Police, and Punish the Poor*, (New York: St Martin's), 2018

84   Simonite, Tom 'When It Comes To Gorillas, Google Photos Remains Blind', *Wired.com*, 11 January 2018 <https://www.wired.com/story/when-it-comes-to-gorillas-google-photos-remains-blind/> (accessed on 2 December 2018)

85   Vincent, James "Welcome To The Automated Warehouse Of The Future" *TheVerge,com*<https://www.theverge.com/2018/5/8/17331250/automated-warehouses-jobs-ocado-andover-amazon> (accessed on 20 January 2019)

86   *Statista* 'Revenues From The Artificial Intelligence (AI) Market Worldwide From 2015 to 2024 (In Billion U.S. Dollars', 2018 https://www.statista.com/statistics/621035/worldwide-artificial-intelligence-market-revenue/ (accessed on 3 December 2018)

87   Ransbotham, Sam, Kiron, David, Gerbert, Philipp and Reeve, Martin 'Reshaping Business With Artificial Intelligence: Closing the Gap Between Ambition and Action', *MIT Sloan Management Review*, 6 September 2017

88   Brynjolfsson, Erik and McAfee, Andrew 'The Business of Artificial Intelligence: What It Can—And Cannot—Do For Your Organization', *Harvard Business Review*, 26 July 2017

89   Lynch, Shana 'Andrew Ng: Why AI Is the New Electricity', *Insights By Stanford Business*, 11 March 2017 <https://www.gsb.stanford.edu/insights/andrew-ng-why-ai-new-electricity> (accessed on 3 December 2018)

90   McAfee, Andrew and Brynjolfsson, Erik *Machine, Platform, Crowd: Harnessing Our Digital Future*, (New York: W. W. Norton & Company, 2017)

91 Beck, Andrew *Artificial Intelligence For Computational Pathology*, 15 March 2017 <https://mlhc17mit.github.io/slides/lecture6.pdf> (accessed on 4 December 2018)

92 We would like to thank our friend and colleague Hajime Hotta for this idea.

93 ImageNet *ImageNet Large Scale Visual Recognition Challenge (ILSVRC)*, 2017 <http://image-net.org/challenges/talks_2017/ILSVRC2017_overview.pdf> (accessed on 3 December 2018)

94 Meeker, Mary 'Internet Trends Report 2017', *KPCB*, 31 May 2017 <https://www.slideshare.net/kleinerperkins/internet-trends-2017-report> (accessed on 3 December 2018)

95 Agrawal, Ajay, Gans, Joshua, and Goldfarb, Avi *Prediction Machines: The Simple Economics of Artificial Intelligence* (Boston: Harvard Business School Press, 2018)

96 Agrawal, Ajay, Gans, Joshua, and Goldfarb, Avi *Prediction Machines: The Simple Economics of Artificial Intelligence* (Boston: Harvard Business School Press, 2018)

97 Danziger, Shai, Levav, Jonathan and Avnaim-Pesso, Liora 'Extraneous Factors In Judicial Decisions', *Proceedings of the National Academy of Sciences*, 26 April 2011, 108(17), 6889–6892

98 Li, Kai-Fu *AI Superpowers: China, Silicon Valley, and the New World Order* (Boston: Houghton Mifflin Harcourt Publishing Company, 2018)

99 Again, a big thanks to Hajime Hotta for the suggestion

100 Minsky, Carly 'Kai-Fu Lee: No Hope For Europe's Artificial Intelligence Sector', *Sifted*, 14 December 2018 <https://sifted.eu/articles/interview-kaifu-lee-artificial-intelligence/> (accessed on 3 December 2018)

101 New York Times https://www.nytimes.com/2017/06/16/business/dealbook/amazon-whole-foods.html (accessed on Feb 4, 2019)

102 Tse, Terence and Esposito, Mark *Understanding How the Future Unfolds: Using DRIVE to Harness the Power of Today's Megatrends* (UK: Lioncrest, 2017)

103 CB Insights *On Earnings Calls, Big Data Is Out. Execs Have AI On The Brain*, 30 November 2017 <https://www.cbinsights.com/research/artificial-intelligence-earnings-calls/> (accessed on 4 December 2018)

104 Fry, Hannah *Hello World: How to Be Human in the Age of the Machine*, (London: Doubleday, 2018)

105 "MMC Ventures (2019) The State of AI 2019: Divergence"

106 There are more than a few people who have made this claim, including CEOs, scientists, and at least one European Commissioner. See Haupt, Michael '"Data is the New Oil"—A Ludicrous Proposition', *Medium*, 2 May 2016 https://medium. com/project-2030/data-is-the-new-oil-a-ludicrous-proposition-1d91bba4f294 (accessed on 4 December 2018)

107 Hern, Alex 'Why Data Is The New Coal', *The Guardian*, 27 September 2016 <https://www.theguardian.com/technology/2016/sep/27/data-efficiency-deep-learning> (accessed on 4 December 2018)

108 Charette, Robert N. 'Samsung Securities' $105 Billion Fat-Finger Share Error Triggers Urgent Regulator Inquiry', *IEEE Spectrum*, 13 April 2018 <https:// spectrum.ieee.org/riskfactor/computing/it/samsung-securities-105-billion-fatfinger-share-error-triggers-urgent-regulator-inquiry> (accessed on 4 December 2018)

109 Canny, William 'Deutsche Bank's Bad News Gets Worse With $35 Billion Flub', *Bloomberg*, 19 April 2018 <https://www.bloomberg.com/news/ articles/2018-04-19/deutsche-bank-flub-said-to-send-35-billion-briefly-out-the-door> (accessed on December 4, 2018)

110 The three considerations are drawn from Goldfein, Jocelyn and Nguyen, Ivy 'Data Is Not The New Oil', *Techcrunch*, 27 March 2018 <https://techcrunch. com/2018/03/27/data-is-not-the-new-oil/> (accessed on 5 December 2018)

111 Hernandez, Marco and Duhalde, Marcelo 'How The Asian Face Got Its Unique Characteristics', *South China Morning Post*, 20 February 2019, <https://www. scmp.com/infographics/article/2100532/how-asian-face-got-its-unique-characteristics> (accessed on 20 February 2019)

112 Statista. 'Number of Apps Available in Leading App Stores as of 3rd Quarter 2018', <https://www.statista.com/statistics/276623/number-of-apps-available-in-leading-app-stores/> (accessed January 23, 2019)

113 Freier, Anne, 'App Revenue Reaches $92.1 Billion In 2018 Driven By Mobile Gaming Apps', 13 Sept 2018, Business of Apps <http://www.businessofapps.com/ news/app-revenue-reaches-92-1-billion-in-2018-driven-by-mobile-gaming-apps/> (accessed January 23, 2019)

114 Alter, Lloyd 'People Are Outraged To See Refugees With Smartphones. They Shouldn't Be', *MNN.com*, 8 September 2015, <https://www.mnn.com/green-tech/ gadgets-electronics/blogs/people-are-outraged-see-refugees-smartphones-they-shouldnt-be> (accessed on January 24, 2019)

115 Mosher, Dave, and Kiersz, Andy, 'Elon Musk Has Job Openings For More Than 500 People At Spacex—Here's Who The Rocket Company Wants To Hire', *Business Insider*, 7 Jun 2018 <https://www.businessinsider.com/spacex-jobs-elon-musk-hiring-open-positions-2018-6> (accessed January 23, 2019)

116 Hiner, Jason, 'AI Will Eliminate 1.8M Jobs But Create 2.3M By 2020, Claims Gartner', *Tech Republic*, 2 October 2017, <https://www.techrepublic.com/ article/ai-will-eliminate-1-8m-jobs-but-create-2-3m-by-2020-claims-gartner/> (accessed January 24, 2019)

117 "Mandate for the International Panel on Artificial Intelligence", Justin Trudeau, Prime Minister Of Canada, 6 December 2018, <https://pm.gc.ca/eng/ news/2018/12/06/mandate-international-panel-artificial-intelligence> (accessed January 24, 2019)

118 Government.ae, 'UAE Strategy for Artificial Intelligence', Official Portal of the UAE Government, October 2017 <https://government.ae/en/about-the-uae/ strategies-initiatives-and-awards/federal-governments-strategies-and-plans/ uae-strategy-for-artificial-intelligence > (accessed January 25, 2019)

119 Malek, Caline, 'UAE Sees Future In Artificial Intelligence', *Arab Weekly*, 12 Sept 2018, <https://thearabweekly.com/uae-sees-future-artificial-intelligence> (accessed January 25, 2019)

120 BBC News, 'Chinese Man Caught By Facial Recognition At Pop Concert', 13 April 2018, <https://www.bbc.com/news/world-asia-china-43751276> (accessed 24 January 2019)

121 Russell, Jon, 'China's CCTV Surveillance Network Took Just 7 Minutes To Capture BBC Reporter', *TechCrunch*, 10 Dec 2017, https://techcrunch. com/2017/12/13/china-cctv-bbc-reporter/ (accessed January 24, 2019)

122 BBC News, 'Chinese Man Caught By Facial Recognition At Pop Concert', 13 April 2018, <https://www.bbc.com/news/world-asia-china-43751276> (accessed 24 January 2019)

123 Zhou, Jiaquan, 'Drones, Facial Recognition And A Social Credit System: 10 Ways China Watches Its Citizens', *South China Morning Post*, 4 August 2018 <https:// www.scmp.com/news/china/society/article/2157883/drones-facial-recognition- and-social-credit-system-10-ways-china> (accessed January 24, 2019)

124 MacLean, Asha, 'e-Estonia: What is All the Fuss About?', *ZDNet*, 13 August 2018 <https://www.zdnet.com/article/e-estonia-what-is-all-the-fuss-about/> (accessed January 25, 2019)

125 Kissinger, Henry, 'How the Enlightenment Ends', *The Atlantic,* June 2018 <https://www.theatlantic.com/magazine/archive/2018/06/henry-kissinger-ai- could-mean-the-end-of-human-history/559124/> (accessed January 25, 2019)

126 Cadwalladr, Carole and Graham- Harrison, Emma, 'Revealed: 50 million Facebook Profiles Harvested For Cambridge Analytica In Major Data Breach,' *The Guardian*, 17 March 2018, https://www.theguardian.com/news/2018/mar/17/ cambridge-analytica-facebook-influence-us-election (accessed on February 1, 2019)

127 Goos, Maarten, Manning, Alan and Salomons, Anna 'Explaining Job Polarization: Routine-Biased Technological Change and Offshoring' *American Economic Review*, 104(8), 2509–26, 2014

128 Harari, Yuval Noah, *Homo Deus: A Brief History of Tomorrow* (New York: Harper Collins, 2017)

129 Giles, Martin 'Artificial Intelligence Is Often Overhyped—And Here's Why That's Dangerous', MIT Technology Review, 13 September, 2018 <https://www. technologyreview.com/s/612072/artificial-intelligence-is-often-overhypedand-heres-why-thats-dangerous/> (accessed on January 25, 2019)

130 Jordan, Michael 'Artificial Intelligence—The Revolution Hasn't Happened Yet', *Medium*, 18 April, 2018 <https://medium.com/@@mijordan3/artificial-intelligence-the-revolution-hasnt-happened-yet-5e1d5812e1e7> (accessed in January 25, 2019)

131 Topham, Gwyn 'Philip Hammond Pledges Driverless Cars By 2021 And Warns People To Retrain', *The Guardian*, 23 November 2017 https://www.theguardian. com/world/2017/nov/23/philip-hammond-pledges-driverless-cars-by-2021-and-warns-people-to-retrain (accessed on January 25, 2019)

132 Fry, Hannah *Hello World: How to Be Human in the Age of the Machine*, (London: Doubleday, 2018)

133 Davidson, Cathy N, *Now You See It: How Technology and Brain Science Will Transform Schools and Business for the 21st Century* (New York: Penguin Books, 2012)

134 Fishman, Tiffany Dovey and Sledge, Linsey, 'Reimagining Higher Education: How Colleges, Universities, Businesses, and Governments Can Prepare For A New Age Of Lifelong Learning', *Deloitte Insight*, 22 May 2014 <https://www2. deloitte.com/insights/us/en/industry/public-sector/reimagining-higher-education.html> (accessed on Jan 25, 2019)

135 Friedman, Thomas L. *Thank You for Being Late: An Optimist's Guide to Thriving in the Age of Accelerations: Pausing to Reflect on the Twenty-First Century*, (London: Penguin, 2017)

136 Centre for the New Economy and Society, 'The Future of Jobs Report', 2018, World Economic Forum, <http://www3.weforum.org/docs/WEF_Future_of_ Jobs_2018.pdf> (accessed January 25, 2019)

137 McGowan, Heather 'STOP Asking What...and Start asking WHY', *Linkedin*, 10 May 2017 <https://www.linkedin.com/pulse/stop-asking-what-heather-mcgowan/> (accessed January 26, 2019)

138 Christensen, Clayton M., Allworth, James, and Dillon, Karen, *How Will You Measure Your Life?* (New York: HarperCollins, 2012)

139 Friedman, Thomas L. *Thank You for Being Late: An Optimist's Guide to Thriving in the Age of Accelerations: Pausing to Reflect on the Twenty-First Century*, (London: Penguin, 2017)

140 Rosen, Larry, and Samuel, Alexandra, 'Conquering Digital Distraction', *HBR.org*, June 2015, <https://hbr.org/2015/06/conquering-digital-distraction> (accessed January 26, 2019)

141 Weller, Chris "Silicon Valley Parents Are Raising Their Kids Tech-Free—And It Should Be A Red Flag". *Business Insider*, 18 February, 2018 <https://www.businessinsider.com/silicon-valley-parents-raising-their-kids-tech-free-red-flag-2018-2?r=US&IR=T> (accessed January 27, 2019)

142 Frank, Terri 'E-books versus Print: Which do we Retain Better?', *DIY MFA*, 27 Nov 2017, <https://diymfa.com/reading/e-books-versus-print-retain-better> (accessed January 27, 2019)

143 Cox, Stefani, 'China Is Teaching Kids to Code Much, Much Earlier than the U.S.', Big Think, 19 Nov 2015, <https://bigthink.com/ideafeed/china-is-already-teaching-coding-to-the-next-generation> (accessed January 28, 2019)

144 MacLean, Asha, 'e-Estonia: What is All the Fuss About?', ZDNet, 13 August 2018, <https://www.zdnet.com/article/e-estonia-what-is-all-the-fuss-about/> (accessed January 25, 2019)

145 Euractive, 'INFOGRAPHIC: Coding at school—How do EU countries compare?', 16 Oct 2015, <https://www.euractiv.com/section/digital/infographic/infographic-coding-at-school-how-do-eu-countries-compare/> (accessed January 28, 2019)

146 Kross, Ethan, Verduyn, Philippe, Demiralp, Emre, Park, Jiyoung, Lee, David Seungjae, Lin, Natalie, Shablack, Holly, Jonides, John and Ybarra, Oscar, 'Facebook Use Predicts Declines in Subjective Well-Being in Young Adults', *PLoS ONE* 8(8), 2013

147 Rosen, Aliza "Tweeting Made Easier", *Twitter Blog*, 7 November, 2017 <https://blog.twitter.com/official/en_us/topics/product/2017/tweetingmadeeasier.html> (accessed on January 26, 2019)

148 Martin, Alex 'Startup Takes Stress Out Of Fed-Up Workers' Exit Plans', *Japan Times*, 28 August 2018 <https://www.japantimes.co.jp/news/2018/08/28/business/startup-takes-stress-fed-workers-exit-plans/#.XFrfeLjgpPY> (accessed on January 26, 2019)

149 Talk with Me Baby, 'Word Gap', n.d., <http://www.talkwithmebaby.org/word_gap> (accessed January 26, 2019)

150 Deruy, Emily, 'Georgia's Fight To End the Childhood Word Gap', The Atlantic, 2 Sept 2015, <https://www.theatlantic.com/politics/archive/2015/09/georgias-fight-to-end-the-childhood-word-gap/432702/> (accessed January 26, 2019)

151 Anthony, K.S., 'DARPA Seeks to Make AI Robots More Moral and Less Potentially Psychopathic By Teaching Them Manners', Outer Places, 28 Jan 2019, <https://www.outerplaces.com/science/item/19264-darpa-ai-human> (accessed January 29, 2019)

152 Kiewra, Kenneth A. "How Classroom Teachers Can Help Students Learn and Teach Them How to Learn", *Theory Into Practice*, 41(2), 2002, p71–80

153 Nye, Pauline A. Crooks, Terence J., Powley, Melanie, and Tripp Gail, 'Student Note-Taking Related To University Examination Performance', *Higher Education*, 13(1), 1984, p85–97

154 Fry, Hannah *Hello World: How to Be Human in the Age of the Machine*, (London: Doubleday, 2018)

155 Susskind, Jamie 'Should We Allow Self-Driving Cars To Help People Kill Themselves?', *Wired*, 12 December, 2018 https://www.wired.co.uk/article/self-driving-car-ethics-morality (accessed January 29, 2019)

156 Groth, Olaf, and Nitzberg, Mark, *Solomon's Code: Humanity in a World of Thinking Machines*, (Cambridge, UK: Pegasus, 2018)

www.ingramcontent.com/pod-product-compliance
Lightning Source LLC
Chambersburg PA
CBHW031401180326
41458CB00043B/6563/J